Common Sense in Environmental Management

Common Sense in Environmental Management examines common sense not in theory, but in practice. Jonathan Woolley argues that common sense as a concept is rooted in English experiences of landscape and land management and examines it ethnographically – unveiling common sense as key to understanding how British nature and public life are transforming in the present day.

Common sense encourages English people to tacitly assume that the management of land and other resources should organically converge on a consensus that yields self-evident, practical results. Furthermore, the English then tend to assume that their own position reflects that consensus. Other stakeholders are not seen as having legitimate but distinct expertise and interests – but are rather viewed as being stupid and/or immoral, for ignoring self-evident, pragmatic truths. Compromise is therefore less likely, and land management practices become entrenched and resistant to innovation and improvement. Through a detailed ethnographic study of the Norfolk Broads, this book explores how environmental policy and land management in rural areas could be more effective if a truly common sense was restored in the way we manage our shared environment.

Using academic and lay deployments of common sense as a route into the political economy of rural environments, this book will be of great interest to scholars and students of socio-cultural anthropology, sociology, human geography, cultural studies, social history, and the environmental humanities.

Jonathan Woolley is an Affiliated Researcher at the Department of Social Anthropology at the University of Cambridge, UK. He was awarded his PhD in March 2018, following over a year of ethnographic fieldwork in the Broads National Park, upon which this book is based. Jonathan's research there was part of an AHRC-funded research project at the University – *Pathways to Understanding the Changing Climate* – which explored the styles of learning about the environment that exist in different cultures around the world. Jonathan has also written on East Anglian folklore, nature spirituality, and public engagement with environmental and cultural heritage.

Routledge Explorations in Environmental Studies

Poetics of the Earth
Natural History and Human History
Augustin Berque

Environmental Humanities and the Uncanny
Ecoculture, Literature and Religion
Rod Giblett

Ethical Responses to Nature's Call
Reticent Imperatives
James Magrini

Environmental Education in Indonesia
Creating Responsible Citizens in the Global South?
Lyn Parker and Kelsie Prabawa-Sear

Ecofeminism and the Indian Novel
Sangita Patil

The Role of Non-state Actors in the Green Transition
Building a Sustainable Future
Edited by Jens Hoff, Quentin Gausset and Simon Lex

The Creative Arts in Governance of Urban Renewal and Development
Rory Shand

Common Sense in Environmental Management
Thinking Through English Land and Water
Jonathan Woolley

For more information about this series, please visit www.routledge.com/Routledge-Explorations-in-Environmental-Studies/book-series/REES

Common Sense in Environmental Management

Thinking Through English Land and Water

Jonathan Woolley

Routledge
Taylor & Francis Group
NEW YORK AND LONDON

earthscan
from Routledge

First published 2020
by Routledge
2 Park Square, Milton Park, Abingdon, Oxon OX14 4RN

and by Routledge
52 Vanderbilt Avenue, New York, NY 10017

Routledge is an imprint of the Taylor & Francis Group, an informa business

British Library Cataloguing-in-Publication Data
A catalogue record for this book is available from the British Library

Library of Congress Cataloging-in-Publication Data
A catalog record has been requested for this book

ISBN: 978-0-367-00237-4 (hbk)
ISBN: 978-0-429-40060-5 (ebk)

Typeset in Bembo
by Wearset Ltd, Boldon, Tyne and Wear

DSM
In fulfilment of my vow

Contents

Illustrations

Figures

Tables

Preface

Common sense: a briefing for policymakers

The problem – siloing obstructs effective environmental land management

One of the key challenges faced by policymakers across government is effectively engaging with stakeholders who have fundamentally different perspectives and goals to one another. Due to the privatisation of land ownership and the professionalisation of land management, the Environment Sector is divided into interest groups and professions that struggle to engage with one another, despite relying on shared infrastructure and resources. "Siloing" of this kind inhibits the formation of shared norms and puts strain on the relationships between (i) different stakeholder groups, and (ii) these groups and the branches of government that hold ultimate, statutory responsibility for areas of life with which they are concerned. Disagreement about environmental land management, such as between farmers and conservationists, is therefore more difficult to resolve.

Policymakers should note that prevalent elements of English culture make this problem worse – in particular, the concept of *common sense.*

What is common sense?

"Common sense" is a culturally specific concept in English society, without a precise translation in other European languages. A working definition of common sense is as follows:

> Common sense is a form of ***pragmatic attitude***, developed through daily interactions with the shared material and social context or "taskscape" – that is, the ***commons*** – of a defined group of people.

Common sense is frequently used discursively, particularly to criticise government policies or administrative processes when they do not map onto the expectations entailed by a stakeholder's own idea of what would be "common sense". It underscores *situated pragmatism as a moral cornerstone of English public life.*

How does common sense shape English society and land management?

Common sense encourages English people to tacitly assume that the management of land and other resources should organically converge on a consensus that yields self-evident, practical results. Furthermore, they then tend to assume that *their own position* reflects that consensus. Other stakeholders are *not* seen as having legitimate but distinct expertise and interests – but are rather viewed as being stupid and/or immoral, for ignoring self-evident, pragmatic truths. Compromise is therefore less likely, and land management practices become entrenched and resistant to innovation and improvement.

How should policymakers respond?

- Be aware that common sense encourages English people to trust their own experience over and above evidence-based approaches. Pragmatic arguments may be more persuasive than hard evidence.
- Provide additional funding to employ advocates who can liaise and build trust through face-to-face interactions with stakeholders. This will mitigate the mistrust caused by professional silos, by making the distinct perspectives expressed through these interactions part of day-to-day experience for all concerned.
- Encourage stakeholders to work and communicate frequently with each other and not just with statutory authorities and government – repositioning potential interlocutors as a helpful source of valuable expertise.
- Consider supporting common pool resource management and education initiatives, to mitigate the divisive effects of privatisation of land and professionalisation of expertise.

Acknowledgements

It is with deepest gratitude and respect that I offer my thanks to the land managers and local residents of Broadland, without whose generosity and patience this book would not have been possible. Your wit, wisdom, and insight are the bricks and mortar from which this text is built. I cannot acknowledge most of you by name for the sake of anonymity, but I can offer particular thanks to the staff and volunteers of RSPB Strumpshaw Fen, TCV Norwich, Norfolk Wildlife Trust, and the Broads Authority; Dr Martin George, Meg Amsden, Phil and Jill Wakley, Graham and Nicky Elliott, and Clarke Willis.

I'd like to thank my supervisors, Dr Richard Irvine and Dr David Sneath, for all their advice, support, and guidance. I could not wish for more talented or more kind mentors. I would also like to thank the Division of Social Anthropology, particularly the rest of Cambridge Interdisciplinary Research on the Environment (CIRE), for creating such an inspiring and vibrant context in which to pursue my doctorate.

I would also like to give special thanks to my friends – especially to Farhan Samanani, Hugh Williamson, Beth Singler, Ragnhild Freng Dale, Elizabeth Cruse, Luke Lofthouse, Melissa Demian, Cavin Wilson, Chris West, Richard Robinson, Max Theiler, Craig Burns, Danika Parikh, Jeansun Lee, Gede Fio Parma, Tim Harris, Ariana Power, Amanda Foale-Hart, Shona Goodall, Tom Jaeckel, Marion Messmer, Haydn King, Christine Black, Matt Mahmoudi, Emma Brownlee, David Hawkins, and Harum Mukhayer for their companionship.

I wish to thank my parents Karen and Liam, and my sister Rachel, for the constant love and emotional and financial support they have provided me with; I hope I have done you all proud.

Finally, I would like to thank the Arts and Humanities Research Council, Pembroke College Cambridge, the Mongolia and Inner Asian Studies Unit (MIASU), the Order of Bards, Ovates, and Druids, and my grandparents for their financial support, without which this book would never have been completed.

Introduction

Common-sense questions

While native speakers of English will profess to know common sense when they see it, very few can articulate what common sense *actually is*. Even Thomas Paine's seminal text *Common Sense* (Paine 2012) – one of the most influential works of political theory in the English language – makes no effort to explore the concept after which it is named. "Common sense" simply anchors Paine's central argument: that rebelling against the British Crown was an obvious moral and practical necessity. Paine's treatise is not a discussion of common sense then, but an exercise of it.

This is characteristic of common sense as a concept in use; it is a piece of intellectual furniture, working quietly and communicating much, but always in the background. In this way, common sense carries an uncommon power, and so asking "*What actually is common sense?*" is clearly a valuable exercise. Answering this question involves looking beyond what is taken for granted in English culture: an exercise in what Randall Collins calls "non-obvious sociology" – the kind of social science that is genuinely useful, because it takes us beyond mere description of social facts (Collins 1992). And yet – as will be explored in later chapters – social anthropologists have shown relatively little interest in common sense. While there have been some theoretical discussions of the concept (see Geertz 1983), no anthropologist has ever attempted to examine it ethnographically, as it is put to use by English speakers. This book seeks to do just that.

Of all the tropes and themes invoked by the people I met during my fieldwork, one of the most prominent was common sense. Farmers would voice their frustration over the Common Agricultural Policy by complaining of bureaucrats with no common sense. Conservationists would self-effacingly talk of using reedbeds for flood-mitigation schemes as "simple common sense". Local residents would demand a "common-sense approach" to local development and planning. As with any other anthropogenic environment, deep-seated cultural attitudes like common sense have a direct, though often unacknowledged, role to play in pragmatic decision-making about how the English landscape is designed, shaped, controlled, and used. If you want to fully understand how English culture shapes the English landscape, then understanding common sense has a critical role to play.

Why: why common sense?

My own interest in the question of what common sense meant began long before I began formal study, however. Ever since I was little, I experienced some confusion regarding English social norms because – being dyspraxic – I didn't intuitively understand the subtle cues given through body language. Because English society relies heavily on non-verbal communication, I relied upon instruction provided by my relatives:

> If you want to talk to someone, Jonathan, you can't just walk over and speak at them. If they're already talking to someone else, don't just interrupt! You need to wait until they pause, and make eye-contact with you first.
>
> Remember to bring a bottle of wine when visiting someone's house for dinner.
>
> Ask people about themselves; don't just talk yourself.

While the other children picked up how to be English as they went along, I had to consciously learn this way of life as a system of rules. My learning difficulty ensured that, from an early age, I subjected my social experiences to conscious appraisal. As such, it is fair to say that my own development has ensured that "home" has always been, to an extent, an ethnographic other. Although this research project is based upon a defined period of fieldwork, conducted from August 2014 to September 2015 in the Broads National Park in the British counties of Norfolk and Suffolk, my observations and analysis bear the impressions of growing up as a kind of naïve ethnographer in England.

But in discerning the meaning of common sense for English people, my goal is to do more than merely expand the ethnographic reach of anthropology or indulge my own personal curiosity. As common sense plays a fundamental role in how English people engage with and make sense of their environment, understanding it has significant implications – particularly for rural industries and environmental policy. The argument made below supports a series of policy recommendations, based on the ethnographic insights gained through my fieldwork, and subsequent analysis. These are summarised in the Preface to this book – which is intended as a succinct briefing for policymakers. Time-pressed scholars (or, more likely, students) may also find this section to be a useful precis of the argument contained within these pages, although they may wish to delve into the full chapters to explore my points in greater detail.

This approach – to actively seek to influence the cultural and environmental dynamics about which I write, by explicating a concept that plays a key role within them – runs contrary to a particular trend in Environmental Anthropology in Britain. There is a history of productive ethnographic enquiry in the United Kingdom, including the Mass Observation initiative spearheaded by the anthropologist Tom Harrisson (Harrisson 2009), the essays of Mary Douglas on the significance of food and drink (Douglas 1972, 2003, 3–16), the historical anthropology of Alan Macfarlane (Macfarlane 1978, 1992, 1999), the ethnographic works of Marilyn Strathern on institutions, kinship, and moral life (Strathern 1981, 1992, 2000, 2016), and a collection of monographs, published in *The Sociological Review* earlier this year (see Degnen and Tyler 2017; esp. Irvine 2017). Perhaps of greatest relevance to my own efforts here is Nigel Rapport's classic study of an English village in the Lake District (1993), Strathern's study of the kinship relations in Elmdon in Essex (1981), Frake's work in North Norfolk (1996), and the work of Veronica Strang on the cultural significance of water in Britain (Strang 2004, 2015). Summarising his own project, Frake states that:

> While fully prepared to point out the sometimes deceptive political-economic facades of culturally constructed edifices, this essay is not intended to contribute to the depressing climate of condescending deconstruction that pervades much of current human science, deconstruction which reduces real people to gullible dupes of a hidden malevolence that makes itself visible only to the discerning investigator.
>
> (Frake 1996, 90–92)

Such sentiments are echoed by Rapport, who reminisces that:

> I told my doctoral supervisor that I knew what I wanted to write about: the complexity, the inconsistency of things. Social life was not about neat, mechanical models, about overarching systems, whatever may be the conventional wisdom about structure and function, synthesis and consensus. Social life was farcical, chaotic, multiple, contradictory; it was a muddling-through, which turned on the paradoxical distinction between appearance and actuality.
>
> (Rapport 1993, ix)

My response to critical interventions like this is to ask: If the human sciences are incapable of making critical interventions to cultural life, what exactly are they for? Although oversimplification is to be discouraged, complexity remains, I suggest, the *explanandum*, rather than the *explanans*.

For there are extremely pressing reasons for not ending our analysis with a simple acknowledgement of complexity. We are currently living through a time of ecological, political, and economic crisis – a crisis engendered, as Strang points out, by powerful cultural narratives regarding the "gardening" of the world, which have escalated into an ideology of limitless growth and development that frequently sublimates the problems humanity faces into mere technocratic issues for specialist, top-down bureaucracies to manage (Veronica Strang 2009, 2, 279–281). England's water resources are experienced and imagined in a way that reflects deep cultural themes that have direct impacts upon how water is managed and governed (Veronica Strang 2004, 159). Watery landscapes like the Broads are therefore vital case studies for thinking about what types of cultural underpinnings might engender (or obstruct) truly sustainable ways of living. When policymakers are increasingly looking to the social sciences to understand the socio-cultural, socio-economic, and behavioural factors at play in our current time of crisis, to be content with narrating the chaos and complexity of social life would be blithely complacent at best, and utterly negligent at worst.

This introduction is split into three parts. This, the first, introduces my central objectives. The second introduces the Broads as the fieldsite for my ethnographic work. The third provides a summary of my argument, as it is structured in the following chapters.

Where: the Broads as a fieldsite

The Broads is Britain's largest protected wetland – a lush carpet of rivers, shallow lakes, marshes, and fen, spreading across much of Eastern Norfolk and extending into the neighbouring county of Suffolk (see Figure I.1). The region gains its name from more than 60[1] shallow lakes – the eponymous broads – that dot the landscape, creating a habitat for thousands of different kinds of waterbirds, plants, microorganisms, amphibians, mammals, and fish – many of which are nationally or internationally scarce. These broads are also a popular site for recreational pursuits: from sailing to fishing, from motorboating to painting. The watershed that feeds the floodplains and broads themselves extends far inland and includes some of Britain's richest farmland. The Broads National Park, formally referred to as the Broads Executive Area, is a jurisdiction of some 303 square kilometres, including the lower flood plains of seven interconnected rivers: the Yare, the Bure, the Waveney, the Thurne, the Ant, the Chet, and the Wensum. The topography is gentle, ensuring that the 200 kilometres of inland waterways here have no locks or weirs, allowing for ease of travel. Both navigation and conservation are the responsibility of the Broads Authority, a statutory government agency set up by a UK parliamentary act in 1988, its dual responsibility for conservation and navigation setting it apart from other authorities in the National Parks family (see Figure I.2). In addition to maintaining the rivers for navigation, and safeguarding the unique habitats found there, the Broads Authority encourages tourism, oversees planning, liaises with landowners to ensure water quality, and helps coordinate the activities of non-governmental organisations (NGOs), community groups, and private businesses that operate in the Broads. Most of the local population live on higher ground that lies outside of the Broads Executive Area, beyond the direct oversight of the Broads Authority. This densely populated and intensively farmed "upland" nonetheless falls within the catchment of Broadland's rivers, and forms a crucial part of the overall landscape of the region.

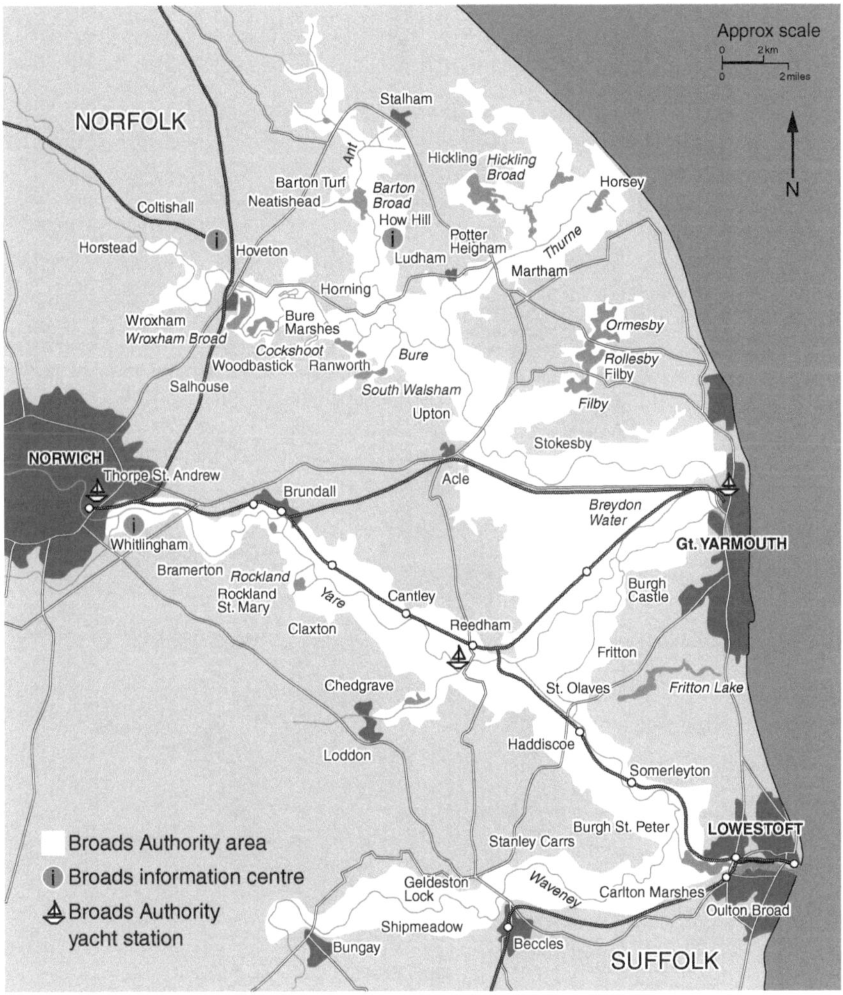

Figure I.1 A map of the broads, towns, and rivers within the Executive Area. The majority of my fieldwork was conducted in the watershed of the Yare (between Norwich and Reedham), the Waveney (around Beccles), and the Bure.

Source: the Broads Authority (2017)

But despite being topographically open – wide and flat, with few trees, at least historically – Norfolk has a reputation for being socially closed. In 1927, Henry Vollam Morton, a pioneering travel writer, remarked that "Norfolk is the most suspicious county in England" (2000, 231). Ninety-odd years later, it appears little has changed. People would often remark to me that, in comparison to other parts of Britain to the North and West, Norfolk was a very private,

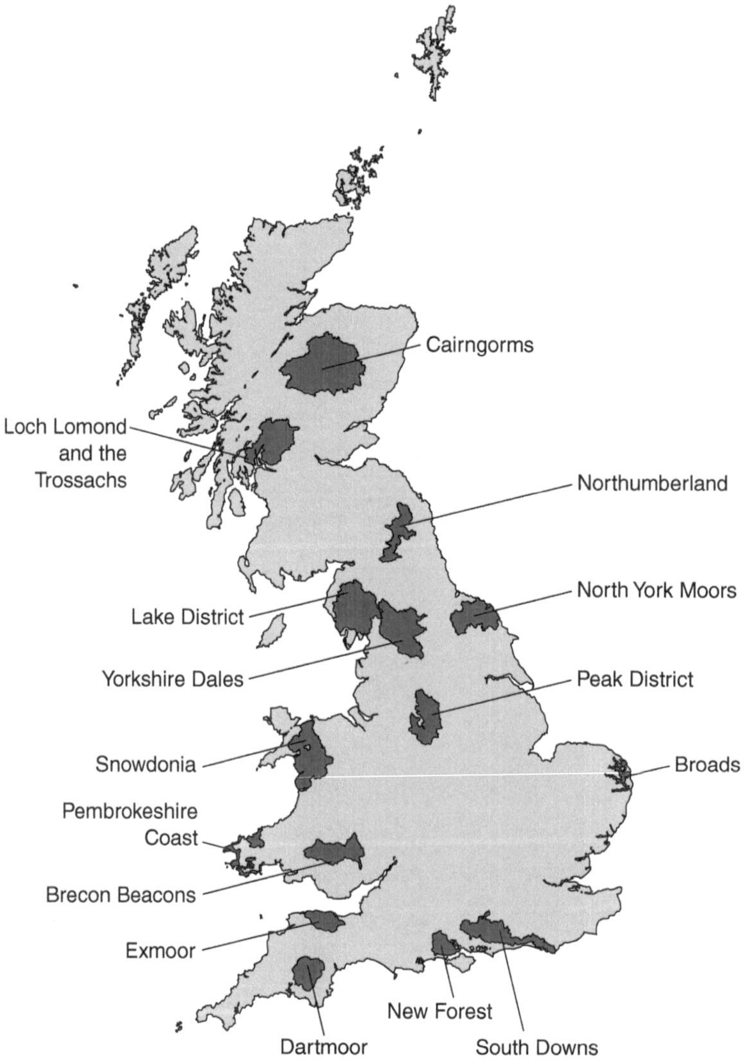

Figure I.2 A map showing the UK's National Parks.
Source: National Parks UK (2016)

individualistic place – where there was little desire to reach out to one's neighbours. One friend of mine struck a stark contrast between Norfolk's strictly private quality, and his experience growing up on Orkney, where Orcadians would invite passers-by into their homes for tea, and "you'd have a plate of neeps and tatties in front of you before you knew what was happening!" Such a social context would, naturally, have been an easier place to do long-term participant observation. In Norfolk, although people were very happy to help me with my research, the relationship they sought to develop was – in many cases

– confined to a brief interview at their place of work.[2] There was little question of inviting a stranger into one's house for a full meal, much less so adopting them into your family or social group. As such, although I collected a large number of semi-structured interviews, I was at first unable to conduct much in the way of participant observation. As we shall see, this problem was compounded by the fact that most local people don't actually live in the Broads proper. Ironically, what resolved this issue of not being able to break into people's personal lives with the Broads as my object was securing full-time employment. Once I started working as a volunteer five days a week on smallholdings or at RSPB Strumpshaw Fen, I got to know people as friends. Being a colleague was a culturally salient reason to accept me personally, whereas the status of an anthropologist interested in the local culture was not.

A key rhetorical function of ethnography is to convey the "wholeness" of a given society to the reader (Thornton 2013). And yet the reality of life in Broadland is anything but holistic, as my difficulties during fieldwork amply illustrate. What follows is therefore a testament to the fractured nature of social life in Broadland; rather than providing a holistic ethnographic portrait of the region, it is rather a series of conversational interviews, stitched together by the work experience I gained in and around the Park.

The Broads has not previously been subject to much formal anthropological study. M. L. Braddy's 2002 PhD dissertation on rural Methodism involves an ethnographic component (Braddy 2002; Stringer 2011, 56–59), but Braddy was working under the auspices of theology, rather than anthropology. Charles Frake paints a vivid portrait of the constitution of local identity through perceptions of place in his essay *A Church Too Far Near a Bridge Oddly Placed: The Cultural Construction of the Norfolk Countryside* (Frake 1996). Nonetheless, the Broads has been subject to an affectionate popular literary discourse – alas "much of it is repetitive, adulatory, rose-coloured and unsatisfying for anyone wishing a real understanding" (Moss 2001, 24). *In the Nature of Landscape* (2014) by the cultural geographer David Matless surveys this canon, alongside the many other varied narrations about Broadland produced by tourists, the press, academics, and local people. Matless's study of Broadland culture is exhaustive, and highly ethnographic in detail, comparable to Frake's brief, but rich portrait of the Broads (Frake 1996, 96–97). Matless's thorough survey of the expert and touristic literature, I feel, leaves little to be added from that perspective. Matless is himself a Broads enthusiast, and it is this perspective that dominates *In the Nature of Landscape* – simply because enthusiasts produce the most writing about the Broads. This analysis of Broadland narratives is a worthy enterprise, but it leaves open questions regarding political economy and socio-ecology; how does the distinctive culture and environment of Broadland relate to flows and accretions of labour, material, and power? My own experiences – based on work, rather than leisure – are well placed to explore such themes.

By focusing on the working life of this landscape, my research followed in the footsteps of scholars in other fields, primarily naturalists, ecologists, landscape historians, and geographers. Given the fact that the Broads is an anthropogenic landscape, many naturalists – particularly those working in the late nineteenth and early twentieth century – documented the human elements to the landscape too. At times, these accounts enter a distinctly ethnographic mode, reminiscent of the early work of Franz Boas (Boas 1964). This genre of Broadland near-ethnography is perhaps exemplified by *The Norfolk Broads* (1903) by William A. Dutt (1870–1939), a journalist based in Lowestoft and London, who wrote extensively on the countryside of East Anglia. *The Norfolk Broads* provides a vivid description of the Broadland landscape, complete with beautiful illustrations of rural scenes and wildlife. Dutt's knowledge was founded upon time spent talking to local people – the frontispiece acknowledges that Dutt was "assisted by numerous contributors". As Dutt himself says:

> The visitors who content themselves with what they can see of Broadland from a yacht's deck can never become really acquainted with the Broads and Broadland life. To gain a real knowledge of these, they must, to some extent, "rough it" as the early adventurers did: trudge the river walls; associate with the eel-catchers, marshmen, reed-cutters, and Breydon gunners; explore the dykes unnavigable by yachts, and the swampy rush marshes where the lapwings and redshanks nest; spend days with the Broadsman in his punt, and nights with the eel-catcher in his house-boat; crouch among the reeds to watch the acrobatic antics of the bearded titmice, and fraternise with the wherrymen at the staithes and ferry inns. *If the stranger in Broadland is unwilling to do these things, he must rest content with the outward aspect of the district and second-hand knowledge of its inner life.*
>
> (Dutt 1903, 28, emphasis mine)

Dutt, together with figures like P. H. Emerson (Emerson 1893, 1885), Walter Rye (Rye 1885), and Arthur Patterson (Tooley 1985), built a reputation as "authorities" on the region, supported by long-term participant observation with the rural workforce. In many respects, I modelled my approach on the example set by these scholars – who spent most of their time with those who hunted, farmed, herded, and managed the Broads, rather than enthusiasts who visited the place. Of the most recent generation of writing about the Broads, I take most inspiration from Mark Cocker's *Crow Country* (2008), a deeply affectionate account of the Mid Yare Valley – the region in which I did the majority of my own fieldwork. Cocker, who I was lucky enough to meet during my time in Norfolk, is a naturalist who devotes most of his attention to the rooks who nest at Buckenham Carr, but their foraging flights do lead him to delve into the lives of those human beings who live and work nearby (Cocker 2008, 88–97). Whereas Matless and Frake provide a rigorous portrait of the Broads "from the yacht deck" as Dutt put it, I wished to "rough it" (See Figure I.3).

Figure I.3 From the yacht deck. A typical Broadland vista in the Yare Valley. Although it may look "wild" and free of human interference to visitors, this landscape bears all the hallmarks of intensive management – including removal of riverside trees to allow the wind to fill the sails of yachts, the regular management of the reedbeds (right), and even the straightening and dredging of the Yare itself.

Credit: Author

Although it is often imagined in popular accounts as a wilderness, the Broads is – and has always been – a peopled landscape. Although the same can be said for many supposed "wilderness" areas (Cronon 1997a, 1997b), the Broads perhaps exemplifies this, as its defining features are the products of human artifice. Botanist and ecologist Joyce Lambert used stratigraphic sampling of sediments across the region to conclusively prove – from their unnaturally steep sides – that the broads were artificial in nature: the product of extensive medieval peat-diggings that had flooded in subsequent centuries (Lambert *et al.* 1960; George 1992). Many of the region's best-beloved features – from windpumps to the grazing marshes – betray its industrial character. Rather than having been created through a ruthless and total expropriation of the local population, as in colonial contexts (Rudin 2013; Mason 2014), the history of Broadland and its eventual transition into a protected area could be read as a story of a gradually increasing human presence in the landscape, with increasing numbers of tourists and intensifying agriculture necessitating the creation of a statutory body capable of protecting the region's unique biota (see Ewans 1992). The Broads Executive Area itself is inhabited by over 6,000 people, and a great many large towns and villages such as Acle, Brundall, North Walsham, and Beccles, as well as the conurbations of Norwich, Great Yarmouth, and Lowestoft, lie just outside.

All in all, this makes the Broads an ideal location to study land management ethnographically – the practices and attitudes of land managers are rendered manifest in the peat and plants, wood and water, silt and sky of this anthropogenic region. Perhaps the most dominant of these attitudes – pivotal to my experience of the private, atomistic quality to Norfolk sociality – is individualism, particularly *possessive individualism* (Macpherson 1962; Rose 1994; Strang and Busse 2011), according to which the ownership and management of land in Norfolk is organised. Possessive individualism is realised through a process of *enclosure*, by which economic resources that are owned or managed collectively are physically and legally parcelled up into private assets separately owned by specific individuals. Although most strongly associated with the Parliamentary Acts of Enclosure in the eighteenth century, the process of enclosure has a long and complex history (Frake 1996, 99; Pryor 2011, 465–475). Some parts of the British countryside were never organised into open fields, and enclosure has taken place piecemeal ever since Medieval times (Rackham 1997). Below, I shall argue that this complex story of enclosure can be extended further, as a dynamic process that continues to affect economic life to this day. The social impacts of this process are profound; progressive enclosures deprive English people of the means required to build a shared understanding of how the land should be managed – leaving people trapped in "silos" represented by particular professions, trades, and interest groups. *Siloing* leads to mutual mistrust, even hostility, that has profound ecological implications. These concepts often stand in contrast to an alternative framework, that of the *common*. As opposed to an enclosed landscape of siloed, possessive individuals, the common is a shared, open landscape, in which reasonable individuals use *common sense* to discover and maintain common ground. Each of these key concepts will be developed and explored in the chapters below, each of which is centred upon a particular question.

What: a common-sense argument

Of those groups said to lack basic common sense, academics are among those most frequently mentioned. In Chapter 1, *Do academics have common sense?*, I assess various academic discussions of the concept. I argue here that this literature – though extensive – does not sufficiently engage with the everyday usage of the term by English speakers. Academics frequently define common sense either in terms of the universal human capacity for reason, or as a simple expression for received wisdom. But the etymology of the phrase in English indicates a quite different significance of the phrase, particular to vernacular use. I conclude that it is only through examining the term as it is deployed in everyday situations – through ethnography – that it can be properly understood.

Following this, Chapter 2, *What is common sense?*, provides an ethnographic montage of occasions where common sense is used and what it means in

Broadland. Land managers make recourse to common sense on a daily basis, to discuss the extent, substance, and limits of proper, effective behaviour. By examining these discussions, we can construct a working definition of common sense:

> A *pragmatic attitude*, developed through daily interactions with the shared material and social context – that is, the *commons* – of a defined group of people.

From this definition, the diverse meanings associated with the concept are derived.

In Chapter 3, *Where is common sense to be found?*, I suggest that "the commons" acts as a "root metaphor" in the English imagination, that is "a symbol which operates to sort out experience, to place it in cultural categories, and to help us think about how it all hangs together" (Ortner 1973, 1341). The commons knits together wide-ranging ideas about community, moral behaviour and effective action, interwoven with a shared landscape where work takes place. Evidence for this can be found in the environmental history of Broadland, and the "aesthetic of proximity" of the region's farmers. It is farmers – who inscribe their labours on the landscape and are able to "read" the history of their own efforts and those of others in their surroundings – who hold a habitus of the common that would once have been more widespread, and thus hold a distinct "common sense". This conflicts with the continuing process of enclosure we find in Broadland today, which creates an opposing form of habitus among the majority of the local population.

Chapter 4, *Can you learn common sense?* reflects upon how the habitus of the common is acquired. I provide a thick description of my own experiences volunteering at RSPB Strumpshaw Fen, and contrast this with that of visitors to the landscape. Whereas those who manage the reserve are able to develop a shared "aesthetic of proximity", akin to that of farmers, those tourists engaged in quiet enjoyment view the Broads at a remove. Against O'Riordan's positive characterisation of "Broadland Consciousness" among tourists, I follow Cocker's suggestion that the broads themselves act as a psychological divide, or a mere temporary destination, for most people. Despite the continued role of the common as a theme in English cultural imaginaries, the predominance of enclosure and quiet enjoyment, and their distinctive opposing habitus to that of the common, turns the common from a root metaphor into a dead one.

Having explored the positive affirmations of common sense in the previous chapters, Chapter 5, *Why is common sense so scarce?*, deals with a curious fact – that English people often joke that common sense isn't very common. In the context of such a contested cultural space as the Broads, common sense gains the varied shades of meaning it shows in vernacular conversation – from a basic principle that unites all right-thinking people, to a virtue that few possess. This highlights the widespread frustration, even exasperation with the views of others, that characterises public life in the Broads National Park. Such claims serve to demonstrate a central tension that emerges in modern Broadland – where a highly individuated landscape and social milieu conflict with the expectation that society should be organically unified with place. The space between expectation and reality leads to mistrust, which in turn engenders something that Gillian Tett helpfully characterises as "the silo effect" in her recent discussion of corporate cultures and institutional failure (Tett 2016). As this mistrust drives further demand for individuated management of the landscape, and thus further enclosure, the silo effect has become a vicious cycle.

In Conclusions, *What do we need to know about common sense?*, I reflect upon how Durkheim's promise – that the modern division of labour will yield a state of organic solidarity – is broken in the contemporary English countryside. A greater professionalisation of knowledge, specialisation of work, and privatisation of land creates an atmosphere of anomie and mistrust that prevents solidarity networks from forming and frustrates the pragmatic aspirations inherent in the concept of the common. This dynamic has had profound ecological and political consequences – from Brexit, to the continued inaction over the climate crisis. The replacement of common sense with a silo mentality represents an enormous challenge for policymakers seeking to transform our way of life, and regenerate the landscape. If environmental land management is to be implemented successfully in coming decades, common sense must be restored in the ways we manage our shared environment.

Notes

1 This curiously vague figure reflects the fact that these shallow lakes are prone to being colonised by marginal vegetation, and landowners will on occasion cut and dredge entirely new areas of open water; the number of broads, therefore, is constantly changing. Furthermore, there is no agreement of what distinguishes a broad from a pond, other than convention, and many broads form part of the course of rivers or are separated from one another by only narrow strips of land. As such, most authorities decline to pin down the precise figure.

2 People who worked "at home" – such as artisan food producers or large landowners – would invite me into their kitchens and sitting rooms for coffee or tea; but it was clear from the tone and framing of such meetings that this did not constitute an invitation to build a more-long-term friendship.

References

Boas, Franz. 1964. *The Central Eskimo*. S. l. Lincoln: University of Nebraska Press.

Braddy, Malcolm Lorance. 2002. "Religion and Faith in a Norfolk Village". Unpublished PhD Thesis. Birmingham: University of Birmingham.

Broads Authority. 2017. "Who We Are". The Broads Authority. 2017. www.broads-authority.gov.uk/broads-authority/who-we-are#map.

Cocker, Mark. 2008. *Crow Country*. London: Vintage.

Collins, Randall. 1992. *Sociological Insight: An Introduction to Non-Obvious Sociology*. 2nd edition. New York, NY: Oxford University Press.

Cronon, William. 1997a. "The Trouble with Wilderness; or, Getting Back to the Wrong Nature". In *Uncommon Ground: Rethinking the Human Place in Nature*, edited by W. Cronon, 69–90. New York: W. W. Norton & Company.
Cronon, William. 1997b. *Uncommon Ground: Rethinking the Human Place in Nature*. New York: W. W. Norton & Company.
Crown Copyright. 2016. National Parks UK. "National Parks – Britain's Breathing Spaces". Hosted by Ace Geography. Sourced from: www.acegeography.com/uploads/1/8/6/4/18647856/846752_orig.gif.
Degnen, Cathrine, and Katharine Tyler. 2017. "Bringing Britain into Being: Sociology, Anthropology and British Lives". *The Sociological Review Monographs* 65 (1): 20–34. https://doi.org/10.1177/0081176917693494.
Douglas, Mary. 1972. "Deciphering a Meal". *Daedalus* 101 (1): 61–81.
Douglas, Mary. 2003. "A Distinctive Anthropological Perspective". In *Constructive Drinking: Perspectives on Drink from Anthropology*, 3–16. London: Psychology Press.
Dutt, William. 1903. *The Norfolk Broads*. London: Methuen & Co.
Emerson, Peter Henry. 1885. *Life and Landscape on the Norfolk Broads*. Photographic Collection. The Metropolitan Museum of Art [The Met Museum]. www.metmuseum.org/art/collection/search/290484.
Emerson, Peter Henry. 1893. *On English Lagoons*. London: David Nutt.
Ewans, Martin. 1992. *The Battle for the Broads: A History of Environmental Degradation and Renewal*. Lavenham, SFK: Terence Dalton Ltd.
Frake, Charles. 1996. "A Church Too Far Near a Bridge Oddly Placed: The Cultural Construction of the Norfolk Countryside". In *Redefining Nature : Ecology, Culture and Domestication*, edited by R. F. Ellen and Katsuyoshi Fukui. Oxford: Berg.
Geertz, Clifford. 1983. *Local Knowledge*. New York: Basic Books.
George, Martin. 1992. *Land Use, Ecology and Conservation of Broadland*. Chichester: Packard.
Harrisson, Tom. 2009. *Britain Revisited*. London: Faber and Faber.
Irvine, Richard D. G. 2017. "Anthropocene East Anglia". *The Sociological Review Monographs* 65 (1): 154–170. https://doi.org/10.1177/0081176917693745.
Lambert, J., J. N. Jennings, C. T. Smith, C. Green, and J. N. Hutchinson. 1960. *The Making of the Broads: A Reconsideration of Their Origin in the Light of New Evidence*. Royal Geographic Society Research Series 3. London: Royal Geographical Society and John Murray.
Macfarlane, Alan. 1978. *The Origins of English Individualism: The Family, Property and Social Transition*. Oxford: Basil Blackwell.
Macfarlane, Alan. 1992. "On Individualism". *Proceedings of the British Academy* 82: 171–199.
Macfarlane, Alan. 1999. *Witchcraft in Tudor and Stuart England*. 2nd edition. London: Routledge.
Macpherson, C. B. 1962. *The Political Theory of Possessive Individualism: Hobbes to Locke*. 1st edition. Oxford: Clarendon Press.
Mason, Courtney Wade. 2014. *Spirits of the Rockies: Reasserting an Indigenous Presence in Banff National Park*. Toronto: University of Toronto Press.
Matless, David. 2014. *In the Nature of Landscape: Cultural Geography on the Norfolk Broads*. 1st edition. Chichester, West Sussex; Malden, MA: Wiley-Blackwell.
Morton, H. V. 2000. *In Search of England*. London: Methuen Publishing Ltd.
Moss, Brian. 2001. *The Broads: The People's Wetland*. 1st edition. London: Collins.

Ortner, Sherry B. 1973. "On Key Symbols". *American Anthropologist* 75 (5): 1338–1346.

Paine, Thomas. 2012. "Common Sense". In *Political Writings – Cambridge Texts in the History of Political Thought*, edited by E. Kuklick. Cambridge: Cambridge University Press.

Pryor, Francis. 2011. *The Making of the British Landscape: How We Have Transformed the Land, from Prehistory to Today*. London: Penguin.

Rackham, Oliver. 1997. *The History of the Countryside: The Classic History of Britain's Landscape, Flora and Fauna*. 2nd edition. London: Phoenix Giant.

Rapport, Nigel. 1993. *Diverse World-Views in an English Village*. Edinburgh: Edinburgh University Press.

Rose, C. 1994. *Property and Persuasion: Essays on the History, Theory, and Rhetoric of Ownership*. Boulder: Westview.

Rudin, Ronald. 2013. "Removing the People: The Creation of Canada's Kouchibouguac National Park". *Arcadia*, Rachel Carson Center for Environment and Society, no. 17. www.environmentandsociety.org/node/5646.

Rye, Walter. 1885. *The History of Norfolk*. Oxford: Wentworth Press.

Strang, Veronica. 2004. *The Meaning of Water*. Oxford; New York, NY: Bloomsbury Academic.

Strang, Veronica. 2009. *Gardening the World: Agency, Identity and the Ownership of Water*. New York, NY: Berghahn Books.

Strang, Veronica. 2015. "Reflecting Nature: Water Beings in History and Imagination". In *Waterworlds: Anthropology in Fluid Environments: Ethnography, Theory, Experiment*, edited by K. Hastrup, 247–278. New York: Berghahn Books. http://dro.dur.ac.uk/20766/.

Strang, Veronica, and Mark Busse. 2011. "Introduction: Ownership and Appropriation". In *Ownership and Appropriation*, edited by Veronica Strang and Mark Busse, 1–19. Oxford; New York, NY: Berg.

Strathern, Marilyn. 1981. *Kinship at the Core: An Anthropology of Elmdon, a Village in North-West Essex in the Nineteen-Sixties*. Cambridge: Cambridge University Press.

Strathern, Marilyn. 1992. *After Nature: English Kinship in the Late Twentieth Century*. Cambridge: Cambridge University Press.

Strathern, Marilyn. 2000. "Introduction: New Accountabilities". In *Audit Cultures: Anthropological Studies in Accountability, Ethics, and the Academy*, edited by Marilyn Strathern, 1–19. London: Routledge. http://dx.doi.org/10.4324/9780203449721.

Strathern, Marilyn. 2016. "Inroads into Altruism". In *Releasing the Commons: Rethinking the Futures of the Commons*, edited by Ash Amin and Philip Howell, 161–177. London; New York, NY: Routledge.

Stringer, Martin D. 2011. *Contemporary Western Ethnography and the Definition of Religion*. Edinburgh: A&C Black.

Tett, Gillian. 2016. *The Silo Effect: Why Every Organisation Needs to Disrupt Itself to Survive*. London: Abacus.

Thornton, Robert. 2013. "The Rhetoric of Ethnographic Holism". *Cultural Anthropology* 3 (3): 285–303.

Tooley, Beryl. 1985. *John Knowlittle: The Life of the Yarmouth Naturalist Arthur Henry Patterson*. Norwich: Wilson-Poole.

1 Do academics have common sense?

If you treat common sense as you would any other cultural form, the obvious first step would be to consider what other scholars have had to say about the subject. However, in 2001, Jeremy Paxman – a British journalist and political commentator – published words that sound a note of caution about such a seemingly routine step on the road to understanding:

> The English approach to ideas is not to kill them, but to let them die of neglect. The characteristic English approach to a problem is not to reach for an ideology, but to snuffle around it, like a truffle hound, and when they have isolated the core, then to seek a solution. It is an approach which is empirical and reconciling and the only ideology it believes in is Common Sense. The English mind prefers utilitarian things to ideas.
>
> (Paxman 2001, 193–192)

The above quotation is where Paxman attempts to explain the antipathy he perceives, within English society, toward intellectuals, ideology, and idealism in general. Although Paxman goes on to say he does not feel that commitment to this "ideology [of] Common Sense" is the primary cause of this antipathy, his utilitarian "ideology" is treated as a contrasting pole to academic understanding. For Paxman's Englishman, the senses and concrete solutions are privileged over detached reflection. This leaves the reader with two enduring impressions: that the English scorn ideas in favour of practice, and that common sense, insofar as it is an intellectual phenomenon at all, is utterly bound up with that practice. This is a curious characterisation for Paxman to make here, considering that empiricism and utilitarianism are both theoretical schools, favoured by generations of intellectuals, English or otherwise. The work of such academic luminaries as John Locke and David Hume, Jeremy Bentham, and John Stuart Mill, which in truth lies at the very foundations of English quotidian culture and liberal ideology today (Fox 2005; Sober 2008; Ryan 2012), is submerged into a natural matter of perceiving the world in a sensible fashion, and responding to these perceptions in a practical, matter-of-fact way. Although Paxman highlights both the *empirical*[1] and *utilitarian* aspects to English knowledge culture, they could be captured together as *pragmatism*.[2]

Indeed, mere empiricism on its own would presumably be viewed as suspiciously abstract, as conveyed by these excerpts from a book on naval slang: "Common sense, a quality sometimes lacking in university graduates of otherwise high intellect" (Jolly 2000, 112) and "Your average university graduate these days is the sort of bloke who can tell you the square root of a pickle-jar … lid to three decimal places – but can't get the bloody thing off …" (ibid., 327).[3] The meaning here is clear; empiricism needs to be part of a *pragmatic* attitude for it to be common sense. Furthermore, intellectuals – like university graduates – are deemed to lack this attitude, despite their empirical abilities. I would, therefore, argue that rather than conceive of common sense as an English species of empiricism as Paxman invites us to do, it is rather an English *pragmatism* – continuously oriented towards solving practical issues over more abstract questions.

In addition to pointing us towards pragmatism as a defining feature of common sense, Paxman's musings and the humour of naval dictionaries help us in another way too: they highlight the tense relationship between intellectuals and the rest of English society, in which common sense plays a key role. This tension both affirms the value of expertise outside the high hall of academe and gives voice to popular criticism of the people that inhabit those halls. "Academics may be very learned" my informants would say "but they just lack common sense". Reflecting upon this tension – between scholars and the public, and the knowledge they hold – is a helpful place to begin, because before we can truly understand what common sense is, we must understand who uses it, and why. But as we shall see, despite their supposed lack of common sense, intellectuals from numerous disciplines have tirelessly explored the concept. How effective these explorations have been, however, remains to be seen.

Koinē aísthēsis and other opinions: key philosophical debates on common sense

While common sense in English culture may be the pragmatic antithesis of intellectualism, this has not stopped generations of philosophers from using the phrase, and in turn influencing popular conceptions of it. And despite the quintessential "Englishness" of common sense itself, such academic engagements draw heavily on an intellectual lineage that reaches far beyond English shores. As such, "common sense" – as referred to in academic philosophy – denotes a quite distinct set of meanings to those conveyed by its lay usage. Descartes' *bon sens*, Kant's *gesunder Menschenverstand*, Reid's *common sense*, etc., are all different from one another, and from lay usage, in their meanings and broader cultural significance. The dialogue – between distinct intellectual and popular understandings of common sense, between more or less pragmatic, and more or less abstract attitudes – mirrors a dialectical relationship at the heart of English attitudes toward knowledge. The polysemy of common sense both emerges from and sits at the heart of this dialogue. Before the common sense can be examined ethnographically, it is helpful to consider the philosophical tradition against which this very English pragmatism is so often contrasted.

* * *

Academic discussion of common sense reaches back to Ancient Greece, and the writings of Aristotle. In *On the Soul*, Aristotle reasons for the existence of a basic sensory capacity – common to all animals – that is able to integrate perception from the other senses into a single sensorium, a capacity he called *koinē aísthēsis*, literally "the common sense/perception" (Aristotle 1957). In the writing of prominent Roman politicians – including Cicero – *communis sensus* was used to refer to a set of concepts and reasoned ideas that are shared in common by a group of people – specifically, the Roman citizenry (Bugter 1987, 89). Despite the fact that both terms are translated in English as "common sense", the distance between these concepts is striking: both owe a significant legacy to quite different intellectual projects and cultural contexts. Aristotle, for example, was seeking to provide an explanation for the agency shown by non-human animals without attributing them souls and reason, as his mentor Plato did – a position Aristotle rejected. Both Plato and Aristotle were drawing upon ancient medical theories about human physiology – with competing schools claiming that either the brain or the heart was the seat of perception (Gregoric 2007, 10:7). Cicero, meanwhile, was sympathetic towards Stoic philosophy, and so his use of a civic common sense reflects the Stoic agenda of the mapping of nature, reason, and ethics onto one another, in order to create a robust basis for personal behaviour and societal norms (Johnson 2013).

Aristotle and Cicero's models of common sense are part of a legacy bequeathed to many subsequent philosophies, from René Descartes' discourse on sensation and reasoning (Descartes 1709, 1; Wilson 2012, 32–45), to the aesthetics of Kant (Kant 1914, 92–94). Broadly speaking, though, it appears the Greek and Latin literatures have given rise to two distinct traditions. Thinkers drawing upon Aristotle link common sense to the senses and basic cognitive faculties pertaining to them, as propounded by the Scottish Common Sense Realist School of the Enlightenment (see Johnston *et al.* 1915). Those drawing on Cicero and the Stoics treat common sense as simply the shared knowledge base of any given society, a view exemplified by Giambattista Vico (Vico 1948; Bayer 2008). Both traditions, however, stress the universality of the feature to which they refer – either a universal human capacity for sensibleness (Reid 2005) or universal human capacity to develop shared assumptions and ideas (Bourdieu 1977).[4] This universalism is repeated by present-day philosophers when common sense is treated matter-of-factly as something that should – at least in principle – be shared by everyone (Chisholm 1982; Elio 2002; Pollock 2002; Lemos 2009). This universalism has even been transplanted into other disciplines, including psychology (e.g. Bogdan 1991) and anthropology (see below).[5]

These philosophical deployments of "common sense" – in various languages – to discuss the twin universals of reason and the co-creation of culture are a worthy enterprise in their own right, about which much more could be said. But here, it is sufficient to note their role as the root stock onto which the vernacular phrase of "common sense" was grafted. But while that vernacular draws strength from philosophical discourse, as its etymology shows, it has grown in a quite distinct direction.

"Sons of the Soil": etymologies of common sense

Through much of the Middle Ages, common sense was known as "common wit" (Simpson and Weiner 2009). But from the 1500s, we begin to see discussions in English of "the common sense" – "sense" and "wit" conveying identical meaning in Early Modern English. By and large, the understanding of "the common sense" shown at the time was squarely of the Aristotelean sort. But we see the beginnings of something different in the debate between George Joye and William Tyndale over the nature of the resurrection in 1543. Joye remarks that Tyndale cannot be so far from his "common senses" as to believe that the dead can hear the voice of Christ.[6] Here, we see a connection being drawn between sensory perception and reason. The truth, here, is evident in universal sensory experience – seeing is believing. The rhetorical force Joye attributes to the senses has, in the English cultural imagination, been commanded by them ever after. This marriage between thought and sensation gathered momentum after the Reformation, so that in 1744 James Harris remarked: "Common sense … a sense common to all, except lunatics and idiots." At the same time, we see common sense acquiring a distinctive, normative flavour too. In 1726, the poet and political writer Nicholas Amherst produced a speech in which he claimed that:

> There is not (said a shrewd wag) a more uncommon thing in the world than common sense, … By common sense we usually and justly understand the faculty to discern one thing from another, and the ordinary ability to keep our selves from being imposed upon by gross contradictions, palpable inconsistencies, and unmask'd imposture. By a man of common sense, we mean one who knows, as we say, white from black, and chalk from cheese; that two and two makes four; and that a mountain is bigger than a mole-hill: in short, when we say a man has common sense, we only say, he is not a fool which (as uncourtly as it may sound) is a very great character; a character, which most men indeed pretend to, but what very few deserve. For though common sense, as before defined, is what the most vulgar and unlearned think themselves possess'd of; yet is it in the most learned often wanting: we are all born without it, and most of us educated in defiance of it; such obstacles and prejudices lie in its way, that it is attained (if at all) with great struggle, pain, and anxiety;

> and when attained (a melancholy consideration!) it comes accompanied with infamy and contempt.
>
> (Amherst 2004)

Amherst's wry commentary is revealing in two ways. First, it hinges on the ironic claim that common sense is, in fact, quite uncommon – a remark I encountered in the field on numerous occasions, and to which I return in Chapter 5. Common sense is often invoked at times of disagreement, or when someone has made some obvious mistake ("She just has no common sense whatsoever!"; "I mean, it's common sense to wear a coat when it's raining, but he just didn't bother"). On such occasions, it is used to point to the obvious folly of an "other", such as an officious bureaucrat or an unreasonable neighbour, in contrast to the reasonableness of the speaker and the listener. The fact that common sense is often most conspicuous in its absence points to its normative, and even political, significance – being used to exclude some people from communities of shared understanding, while including others. As we have already seen, common sense is still deemed in the popular imagination to be something that intellectuals in general do not possess. Scholars are depicted as more concerned with abstract knowledge about a situation than practical mastery of it. As Amherst explains, common sense instead finds full expression among ordinary English people, and so our attention is drawn to the intimate connection between common sense and social class. Common sense acts as way of creating an authoritative field of vernacular knowledge and policing its membership, serving to exclude intellectual elites, who might otherwise command great power in the realm of ideas. But it is from this confluence – of direct experience, and social norms – that common sense gains its pragmatic character.

Common sense is thus pragmatic in an inherently political way – not least through its power to demarcate communities of shared experience. The continuing significance of this power demarcate who is "in", and who is not, is perhaps reflected in the fact that the primary legislature of the British state is referred to as *the House of Commons*.[7] This exclusive form of "common sense" flies in the face of much of the academic literature described above, which treats "common" as synonym for "universal". The collision between these distinct meanings is described by Anthony Ashley Cooper, Earl of Shaftsbury in 1709. Regarding a discussion between himself and his friends, Ashley Cooper mentions that:

> Among different opinions presented and maintained with great life and ingenuity by various participants, every now and then someone would appeal to "common sense". *Everyone allowed the appeal, and was willing to have his views put to that test, because everyone was sure that common sense would justify him.* But when the hearing was conducted – the issue examined in the court of common sense – no judgment could be given. This, however, didn't inhibit the debaters from renewing the appeal to

> common sense on the next occasion when it seemed relevant to do so. No-one ventured to call the authority of the court into question, until a gentleman whose good understanding had never been brought in doubt very gravely asked the company to tell him what common sense was. He said: *If by the word "sense" we understand opinion and judgment, and by the word "common" we mean what is true of all mankind or of any considerable part of it, it will be hard to discover what the subject of common sense could be! For anything that accords with the "sense" of one part of mankind clashes with the "sense" of another.* And if the content of common sense were settled by majority vote, it would change as often as men change, and something that squares with common sense today will clash with it tomorrow or soon thereafter.
>
> (Anthony Ashley Cooper, Earl of Shaftesbury 1709, 6–7, my emphasis)

On one level, what can be seen here is a collision between the Aristotelean and Ciceronian traditions – between common sense as universal reasonableness, and common sense as the universal human tendency to generate a received wisdom – but there is also something else going on at the beginning, specifically, a tendency to use common sense rhetorically to identify what is both normative and reasonable together, colliding with an ambition to situate that synthesis within all mankind. This ambition, as the gentleman of good understanding intimates, could never be reached, because every people in the world has its own idea of common sense.

Second, many of the examples Amherst uses to illustrate what common sense is – knowing one's chalk from one's cheese, knowing that mountains are bigger than molehills – are themselves vernacular turns of phrase,[8] which all relate back to the landscape of England. "Chalk and cheese" refers to the different landforms of the South and East of England, including Norfolk – the high calcareous downs where only sheep are grazed (the "chalk"), and the rich lowland pastures where cattle are traditionally reared (the "cheese"). Moles are common across England, and it is quite apparent to anyone familiar with her rural landscape that, although annoying, their excavations are not mountainous in scale. In Amherst's remarks we see a fully developed form of the construction that appears in Joye – a confluence of sensible observations, and sound thinking; in which the environment plays a key role. It's perhaps especially fitting that the name of Amherst's position at Oxford University – the *Terrae-filius* – translates into English as "Son of the Soil".[9] The connection between common sense and the landscape, I suggest, has a strong cultural resonance – highlighted by the fact that to speak of "a/the common" in vernacular British English specifically denotes common *land*, rather than other shared spaces or resources.

The earliest recorded incidence of "common[s]" being used to refer to common land in this way, according the OED, dates from the Bury Wills in 1491, which states that: "The northe hede abbutyth vppon the comown of Euston"[10] – referring to the position of a grave (*hede*) adjacent to a piece of common land. The third earliest mention, dating from 1550, refers to the enclosure of that same common. But the notion of a common as involving certain collective rights to land dates back all the way to 1386, and is found in line 100 of Chaucer's *The Merchant's Tale*: "Alle othere manere yiftes hardily, As londes, rentes, pasture, or commune …".[11] One of the oldest references to the word "common" recorded in the OED – in the Cursor Mundi d. 1300 – refers to common rights to land "To pastur commun þai laght þe land/ þe quilk þam neiest lay to hand"[12] – a retelling of the actions of Abraham and Lot in Genesis 13 (Simpson and Weiner 2009). Such textual evidence is indicative of both the antiquity and prominence of land as a species of common resource in England, and shows that "common" and "common land" have been all but synonymous since the Medieval Period. Although other forms of common exist in the English lexicon, this textual evidence highlights that the role of land in "the common" is central, and has been so for centuries. It is through this connection that the pragmatism of common sense becomes rooted in the landscape itself. The contemporary contours of this relationship shall be explored in the following chapters.

Common sense as a social scientific object

With this etymological context in mind, we turn now to the work of linguist Alex Wierzbicka, who has written on the use of "common sense" in contemporary spoken and written English. Wierzbicka examines the febrile confluence of culturally specific meanings and associations that lie behind a variety of terms regularly used in English-language scholarship that, she argues, has important effects upon the shape of philosophy more generally (Wierzbicka 2009). Rather than treat common sense as the English term for a universal human capacity, Wierzbicka argues that "anglophone scholars who write about common sense are often oblivious to the fact that the English phrase *common sense* carries a unique, culture-specific meaning" (ibid., 10:338). Wierzbicka produces the following explication of this culture-specific meaning, as a sequence of postulates:

common sense (approach)

a. it can be like this:
b. someone thinks like this:
c. "something is happening here now
d. I want to do something because of this
e. if I do some things, something bad can happen because of this
f. if I do some other things, ***something bad will not happen because of this***
g. I don't want something bad to happen
h. because of this, I want to think about it for a short time
i. if I think about it for a short time, I can know what I can do"

j. when this someone thinks like this, after a very short time this someone can know it
k. ***(because of this, this someone does something)***
l. it is good if someone thinks like this
m. ***all people can think like this***
n. ***it is good if people do things because they think like this***
o. ***it is bad if someone doesn't think like this.***

(Ibid., 10:351, original emphasis)

What's valuable about Wierzbicka's description of common sense is that she makes explicit a series of components that distinguish the vernacular use of common sense from academic deployments of the phrase. While there is broad agreement that common sense is a pattern of thought (b. h. – j.), the notion that it is geared specifically towards pragmatic problem solving (d. – g.) is quite distinct from the Ciceronian tradition, while both the normative (k. – o.) and spatially situated (a. – c.) elements to the vernacular definition stand quite apart from the universal capacity for sensibleness proposed in Scottish Common Sense Realism, and its Aristotelean forebears. I suggest that, when the normative and ecologically grounded aspects of common sense emerge as clearly from the etymology of the phrase as they do from its vernacular use, it is clear to see how these meanings influence its distinctly pragmatic character.

Wierzbicka's work is nonetheless constrained by her sources – namely, the titles of books and legal decisions from English-speaking countries, and the upper middle class cultural commentary of English-speakers like Paxman and Fox. Such material exists in a state of tension with the pragmatic, "empirical" nature of common sense itself, as we saw with Amherst. The authorship of cultural commentary, and the legal profession that provide Wierzbicka with so much of her material are both occupations that involve abstraction from the sort of immanent sensibleness that Wierzbicka herself acknowledges is the defining attribute of common sense. Although lawyers and publishers undoubtedly aspire to embrace common sense as a value – when producing accessibly titled books and proclaiming reasonable legal judgments – both could be considered abstract professional domains that are attempting to *reflect* common sense, properly located elsewhere, rather than spheres where common sense itself is generated, learned, and experienced in the first instance. While bringing common sense into question by drawing attention to its distinct, vernacular uses, Wierzbicka's analysis only takes us so far – as a linguist, she leaves unexplored the norms and situations where common sense takes place. *Where* is common sense properly situated? *What* actions are defined as "common sense", and which are not – especially in the pejorative sense of those deemed to *lack* "common sense"? This, then, begs the question – what do social scientists and anthropologists say in answer to these questions, where common sense is concerned?

Currently, the ethnographic literature on this subject is remarkably sparse. Most anthropologists who make use of, or refer to common sense at all, treat it as a human universal, in the manner of the philosophers. Although it is sometimes used as a means of invoking universal reason for rhetorical purposes[13] (e.g. McFate and Fondacaro 2008), the usual tack taken is more descriptive. In the manner of Cicero, Vico, and other more recent philosophers (van Kessel 1987; Watts 2011), many other anthropologists tend to implicitly treat common sense as "what the ordinary members of society think and do" – in short, as a synonym for received wisdom[14] (e.g. Nadel 1951, 194–199; Webb *et al.* 2002, xiii; Stoler 2009; Szeman 2015). The implications of this view are profound; particularly in terms of how social science engages with wider society. Agustin Fuentes points out that anthropology and common sense both purport to describe societal norms and their relation to universal human reason, so anthropologists must take it upon themselves to debunk fallacious common-sense claims:

> We all need an effective, and robust, toolkit to interpret this reality. Anthropology can be of assistance.... One must be an active learner and a critical thinker, always. Otherwise, we are doomed to sit back and ride the flow of common sense ... and if you do this, beware of the hidden rocks, the perilous falls around the bend, and the particularly dangerous undertow of complacency.
>
> (Fuentes 2012)

Proctor points out that this same point was made by prominent Indian sociologist Andre Beteille, who cast common sense as social science's primary interlocutor. As Proctor (2012) puts it:

> To give in to the assumption that what is "common sense" (the dominant way to do things) in a group is the *only* way, the natural way, to do things is dangerous and fallacious. Anthropology is an important instrument of generating social clarity in this regard.

For Beteille:

> [Sociology] has a body of concepts, methods and data, no matter how loosely held together, for which common sense of even the most acute and well-informed kind cannot be a substitute. For one thing, sociological knowledge aims to be general if not universal, whereas common sense is particular and localised. Educated, middle-class Bengalis, like other educated or uneducated people anywhere, tacitly assume that their common sense is common sense as such or the common sense of mankind.
>
> (Beteille 1996, 2361)

Sociology, for Beteille, equips us with the intellectual tools to look beyond the reified beliefs of our particular frames of reference, to examine the actual conditions in which people live – an orientation that is, in his words "anti-utopian" (1996, 2365).

Although Proctor, Beteille, and Fuentes' broader point – that anthropologists can critique assumed truths that have damaging consequences – is undoubtedly an important one, it is worth noting that the definition of common sense as "the dominant way to do things" is not subjected to any sustained critique by any of them. This way of looking at common sense owes a significant debt to Clifford Geertz, who goes into greater detail regarding how one can treat common sense as a cultural system, that is, "as a relatively organized body of considered thought, rather than just what anyone clothed and in his right mind knows" (Geertz 1983, 75). Geertz observes that common sense is purported among English speakers to lie in "in-the-grain-of-nature realities", and that "'earthiness' might well have been adduced as another quasi-quality of common sense" (ibid., 93), before subjecting this emic claim to critique. Attempting to prise apart the perception of the world from our attempts to make sense of it, Geertz argues that:

> When we say someone shows common sense we mean to suggest more than that he is just using his eyes and ears, but is, as we say, keeping them open, using them judiciously, intelligently, perceptively, reflectively, or trying to, and that he is capable of coping with everyday problems in an everyday way with some effectiveness. And when we say he lacks common sense we mean not that he is retarded, that he fails to grasp the fact that rain wets or fire burns, but that he *bungles everyday problems life throws up for him*....
>
> (Geertz 1983, 76, emphasis mine)

This, Geertz points out, indicates that common sense cannot be mere perception, but requires some degree of analysis or deliberation, too. For Geertz, common sense is first and foremost a cultural system that mediates naïve experience, at the same time conflating the two. This represents a qualification of the model of common sense proposed by the Ciceronian tradition – although common sense may well exist everywhere, *what constitutes* common sense, Geertz says, varies dramatically according to different cultures:

> If we look at the views of people who draw conclusions different from our own by the mere living of their lives, learn different lessons in the school of hard knocks, we will rather quickly become aware that common sense is both a more problematic and a more profound affair than it seems from the perspective of Parisian café or an Oxford Common Room.
>
> (Ibid., 77)

Michael Herzfeld develops this position further. He agrees with Geertz that various kinds of common sense will be found in every society, and further stresses the extent to which common sense is revealed and obtained through practice, dwelling upon its significance in local power relations, and its ethnographic productivity (Herzfeld 2001). With characteristic aplomb, Herzfeld moves in the last to assert, "We may then prefer to adopt a posture in which the discipline of anthropology becomes, quite simply, the comparative study of common sense" defined, ultimately as "symbolisms in use" (Herzfeld 2001, 2286).

Another intersection between common sense and anthropological theory exists in the study of the senses by environmental anthropologists (Descola and Palsson 1996; Bloch 1998; Howes 2003; Petty 2017). Concerned with exploring the universal and culturally specific dimensions to sensory experience through cross-cultural comparison, scholars working in this field clearly fall closer to the Aristotelean tradition, with its interest in universal cognitive or sensory capacities. Having explored the common features of human experiences of water, Strang concludes that:

> it appears that common human physiological and cognitive processes provide sufficient experiential continuity to generate common undercurrents of meaning.... It seems that meaning is the product not just of human individuals and groups, but *also of the common – and diverse – material characteristics of their environments.*
>
> (Strang 2005, 115, emphasis mine)

Geertz and Herzfeld's commentaries on common sense highlight many of the features I have mentioned above. Geertz helpfully stresses the goal-orientation, and pragmatic "earthiness" of common sense as understood within English parlance, while Herzfeld highlights its normative and political power – a point echoed by Beteille, Proctor, and Fuentes for critical purposes. However, all these texts are somewhat brief, and are not based upon sustained *ethnographic* study of common sense as an emic feature of English society – as such, common sense slips into becoming an analytical construct for academic use that describes aspects of all societies rather than a particular kind of "symbolism in use", found in English-speaking cultures in particular. Wierzbicka's argument – that English-speaking scholars frequently use culturally particular concepts that do a lot of conceptual work – is worth considering here. It's also highly worthwhile to integrate the point – stressed by sensory anthropologists – that material circumstances have a direct impact upon forms of meaning. If we are to truly understand what common sense means in a vernacular English context, direct ethnographic evidence will need to be considered, alongside a direct engagement with the tensions between academic and popular thought.

Common sense as a political object

One of the foremost theorists of the material relationship between elite and folk knowledge is Antonio Gramsci. For anthropology, perhaps the most significant fruit to emerge from Gramsci's corpus is that of hegemony, which has been extensively applied in the study of subaltern societies around the world (e.g. Feierman 1990; Comaroff and Comaroff 1991). But as Donald Kurtz (2013) and Kate Crehan (2002) have argued, many anthropologists have not paid particular attention to the sophistication of this concept in Gramscian thought. Crehan suggests that this is in part due to the popularity in anthropological circles of a succinct summary of hegemony penned by Raymond Williams in *Marxism and Literature* (1977). Williams chooses to set aside generalised power relations pertaining to hegemony in society as a whole – about which both he and Gramsci are elsewhere very clear – in order to focus more fully on its implications for the study of literature.[15] This is a gloss that Crehan and Kurtz claim has led many anthropologists to believe that hegemony is primarily an idealist phenomenon. This narrowing of Gramsci's original formulation, referred to by Crehan as "Hegemony Lite", is quite different to hegemony proper, which includes practical matters as well as abstractions within its purview (Crehan 2002, 175). This can be contrasted with, for example, James Scott's claim that hegemony is simply the word Gramsci gives to the position – originally articulated by Marx and Engels – that the ideas held by the elite occupy a position of total dominance within wider society (Scott 1985, 35; Marx and Engels 2000, 192). Crehan, by contrast, stresses the extent to which Gramsci's prison notebooks are intensely concerned with the "*materiality* of power" (Crehan 2002, 176, original emphasis). Gramscian hegemony, rather than simply reiterating the Marxist notion that the class dynamics dictate ideological patterns, attempts to demonstrate the intensely situated and particular nature of those power relations: how matter and ideas, consent and coercion tessellate uniquely in each and every instance (Gramsci 1971, 12), to have the ultimate effect of reproducing wider inequalities (Silverman 2000). For Gramsci, ideas were inherently tied to the political, and so the divide between thought and material facts was illusory (Gramsci 1971, 326).

It is with this theoretical context in mind that we turn to consider where Gramsci's work addresses common sense. Common sense – or *senso comune* in the original Italian – is for Gramsci the residue left by philosophical introspection in broader social attitudes, lying somewhere between folklore[16] and technical knowledge. Indeed, he states that: "Common sense is the 'folklore' of philosophy" (ibid., 419). Gramsci argues that intellectual effort continually percolates into wider society, informing perceptions and attitudes, gradually becoming "common sense" and, eventually, folklore. But this process is highly situational and inconsistent, with different economic terrains – that is, classes – each giving rise to their own common sense (Gramsci 1971, 330 in Crehan 2002, 115).

As Crehan acknowledges, *senso comune* – despite being a literal translation of "common sense" into Italian – is far less loaded than the English phrase, lacking the "connotations of practical, down to earth good sense in Italian that it does in English" (Crehan 2002, 98). *Senso comune* would perhaps be better translated as "public opinion" or "popular consciousness". Indeed, Italian–English dictionaries opt to translate "common sense" as *buon senso* – literally "good sense" – rather than *senso comune* ("Italian Translation of 'Common Sense'" 2017). *Buon senso* also features in Gramsci's thought; namely as the refined form of common understanding, commensurate with critical philosophy (Gramsci 1971, 325–326). Good sense is knowledge oriented towards the lived experiences of ordinary people, but purged of the limitations and shortcomings that Gramsci observed in *senso comune*.

Gramsci also comments on another key attribute of common sense in the English language, identified above – namely, its contested use among both intellectuals and wider society. According to Gramsci, hegemony is not merely a tool used by elites to subjugate and oppress, but it can also be used by radical actors to contest class dynamics and to generate popular revolutionary sentiment. The crucial role of all intellectuals, for Gramsci, was to foster hegemony to buttress their respective class. Traditional intellectuals – that is, academics, clergymen, and the like – served the interests of the propertied class to which they belonged (ibid., 15). But Gramsci also described the role of "organic intellectuals", who shared the interests of the poor, and developed counterhegemonic thought with popular appeal (Gramsci 1971, 12–13; Crehan 2002, 137–145). David Kurtz summarises the relationship of these organic thinkers to traditional intellectuals, being that of good sense versus common sense: "Common sense to Gramsci refers to traditional conceptions of the world, while good sense implies a coherent culture that is a desirable product of the work of organic intellectuals" (Kurtz 2013, 360). Organic intellectuals, in short, would supplant the *senso comune* with *buon senso*, through their production of counterhegemonic ideas. Organic and bourgeois intellectual factions, for Gramsci, existed in a perennial state of tension with one another; it was their dialectical relationship that would lead to revolution. There is a pragmatic edge to Gramsci's understanding of knowledge, here – *buon senso* is defined by its capacity to support the discrete goal – for the poor – of supporting revolutionary politics, while *senso comune* supports the opposite goal – held by elites – of suppressing such movements.

We find an example of this dialectic in action in the polemical exchange between Thomas Paine and Edmund Burke – two of the key British political voices of the Eighteenth Century. Paine – a man from an ordinary background in the Norfolk market town of Thetford who strove to mobilise the Thirteen Colonies of America against the British crown, argued fervently for liberty and equality, moving to support the French Revolution in turn for similar reasons. Burke, the son of a middle-class Anglo-Irish solicitor, sought to ensure that the Colonies remained part of the British Empire, wishing to maintain the status quo. Although their debate only fully crystallised after Burke spoke out against the French Revolution in his *Reflections on the Revolution in France* (Burke 2005), with Paine countering by publishing *Rights of Man I* and *II* (2012b, 2012c), it is Paine's earlier pamphlet – fittingly entitled *Common Sense* (2012a) – that is more instructive for our purposes here. In *Common Sense*, Paine sought to establish that not only was the cause of American Independence morally right; it was also the most prudent course of action. Early in the text, Paine states that:

> I draw my idea of the form of government from a principle in nature which no art can overturn, viz. that the more simple anything is, the less liable it is to be disordered, and the easier repaired when disordered; and with this maxim in view, I offer a few remarks on the so much boasted constitution of England....
>
> (Ibid., 5)

The text is littered with similar, matter-of-fact invocations, such as:

> Small islands not capable of protecting themselves are the proper objects for government to take under their care; but there is something absurd in supposing a continent to be perpetually governed by an island. In no instance hath nature made the satellite larger than its primary planet; and as England and America, with respect to each other, reverse the common order of nature, it is evident that they belong to different systems. England to Europe: America to itself.
>
> (Ibid., 23)

Burke, too, strikes a similar tone in his speeches to Parliament at the time, seeking to reconcile the dispute and thus keep the colonies part of the British state:

> Again and again, revert to your old principles – seek peace and ensue it; leave America, if she has taxable matter in her, to tax herself. I am not here going into the distinctions of rights, nor attempting to mark their boundaries. *I do not enter into these metaphysical distinctions; I hate the very sound of them.* Leave the Americans as they anciently stood, and these distinctions, born of our unhappy contest, will die along with it....
>
> (quoted in Simms 2008, 142–143)

Paine and Burke's arguments both rest not on abstract reasoning, moral principle, or a refined empiricism – but rather on concrete observations, drawn from the situation, goals, and capacities of the intended audience. Paine's *Common Sense* offers not a *description* of common sense as such, but an *exercise* of it – indeed, that the rightness of independence is mere common sense is the very heart of Paine's argument.

If we apply a Gramscian lens to this debate, we can not only position Paine and Burke very easily within the dialectic of intellectuals, but this dialectic also explains why *Common Sense* was so successful. As an ordinary man, but a man of letters, Paine fits Gramsci's portrait of an organic intellectual, able to use his command of ideas to build a counterhegemony in service of his constituency – the 'ordinary people' of the colonies – thus successfully purging the *senso comune* of the American citizenry of the influence of traditional intellectuals, building instead a *buon senso* that recommended revolutionary action (Paine 2012a, 18, 34–38). The pamphlet was widely circulated, passed around by hundreds of thousands of American citizens over the course of the American Revolution, often read aloud in taverns. Although the Revolution was already underway once *Common Sense* was originally printed, it nonetheless galvanised support for the Republican cause and encouraged men to enlist in its armies. Its impact was highly significant, such that "No other text by a single author can claim to have so instantly captured, and then so permanently held the national imagination" (Ferguson 2000, 465–466). Burke, meanwhile, positioned himself more as a traditional intellectual, using a similar rhetorical style to Paine but with less success. Of course, not all American citizens were truly subaltern – many supporters of independence were virtually aristocratic, in terms of their relationship to the mode of production – and sure enough, more propertied figures like John Adams and George Washington would come to alienate Paine over time (Kramnick and Foot 1987; Kuklick 2012). But this shouldn't distract us from the radical nature of Paine's own personal undertaking, as he saw it – indeed, it is undoubtedly this radicalism that led to his eventual ostracism by the propertied elite of the now-independent colonies.

This brief historical sketch, bringing Gramsci's models into dialogue with Paine and Burke, not only helps us appreciate the trajectories of these two thinkers through the political realities they sought to shape, but it also helps us reframe Gramsci's own terminology to better apply it to the specific culture of the Anglophone world. Common sense – as Paine in particular understood it, and then applied it – is more akin to Gramsci's *buon senso*, than it is to his *senso comune*. Common sense in England should not be understood as mere prevailing opinion – a cultural system that defines the obvious, as Beteille and others would have it – because of its connection with pragmatic goals and actions, as both Herzfeld and Geertz remind us. It is from this, to the English mind, that common sense gains its elevated status – unlike *senso comune*, it has to work.

We find indications of the rootedness of common sense in the pragmatic, material realities of what works – and what Geertz identifies as its "earthiness" – in attempts made to communicate the subtleties of the English concept of common sense in other languages – especially in French, where, Wierzbicka points out, it is possible to speak of *bon sens paysan* ('a farmer's [healthy] *bon sens*') and *bon sens terrien* ('the *bon sens* of [a man of] the soil') (Wierzbicka 2009, 10, 341). These French attempts to articulate what makes

English common sense different from the more familiar *bon sens* of French Cartesianism, point to an important iteration of the word "common", discussed above, that is conspicuously absent from Wierzbicka's account – namely, that of *common land*. Parcels of common land to which local people – *commoners* – had certain economic rights and obligations, possess a heady influence in the English imaginaries surrounding work, practical activity, and knowledge; after all, common sense itself is associated with such traits as being "down to earth" or "grounded"; "earthy" classes of people are thought to be simultaneously practical *and* reliable, as well as being rude and non-intellectual[17] – a construct that evokes the relative status in Medieval English society of "Those Who Worked" (the commoners), compared to the elite, intellectual status of "Those Who Prayed" (the clergy) and "Those Who Fought" (the nobility).[18] Common sense, simply put, is a working attitude.

Conclusion: the need for ethnographies of common sense

I would suggest that bringing in materiality in the form of common land allows us to further develop themes raised by social scientific discussions of common sense. Although common sense is, as Wierzbicka claims, an *attitude* rather than a discrete set of actions, or norms in the vernacular English understanding, it is nonetheless an attitude dependent upon taking action (k.) within a given context (c.) where it has normative force (l. to o.).[19] This connection with normative pragmatic action, bequeaths the concept with a political significance – as underscored by Herzfeld. These political associations often take the form of discussions between different intellectual groups and non-intellectuals, whether we are talking about the friends of the Earl of Shaftesbury and their different common senses; Edmund Burke versus Thomas Paine's organic intellectualism; or between Parisian cafés and Oxford Common Rooms. But as Geertz reminds us, common sense in English also has a pragmatic, "earthy" quality, gained through it being put to use in the fields and factories of the English-speaking world. Given these two things – the importance of sharedness to common sense, and the importance of material, earthy matters – it is likely that common land, and the shared landscape in a broader sense, should be a key site for the development and articulation of common sense as a concept. With this in mind, if a rigorous understanding of common sense is to be achieved, there is a need to move beyond academic discourse, toward an ethnographic examination of common sense in use.

Notes

1 Paxman is not the only popular author writing about English culture to identify empiricism as a key feature of English attitudes toward knowledge; Kate Fox explicitly contrasts the English "stolid, stubborn preference for the factual, concrete and common-sense" with "obscurantist, airy-fairy 'Continental' theorizing and rhetoric" (Fox 2005, 405). In this, they both reflect a broad scholarly tradition

– across anthropology and cultural studies – that identifies empiricism as a distinctive feature of English thought (e.g. Porter 1992, 182; Easthope 2004, viii, 59–114; Parkin 2009, 159).

2 Pragmatism also has its own philosophical school, founded through the work of American philosopher Charles Pierce (Pierce 1905).

3 For example, philosophical empiricism does not claim that the findings of scientific experimentation or experience reveal what is *real*: it merely states that they are "empirically sufficient" (Sober 2008, 129–130). Common sense in English society would not only reject this sort of anti-realism; it would not even recognise the worth in considering it as a possibility. Pragmatism, however, concerned with what "works", would respect the English concern for "utilitarian things".

4 I only touch upon Bourdieu's work here, but I will focus upon the significance of his oeuvre – specifically his *Outline of a Theory of Practice* (1977) and *The Bachelor's Ball* (2008) – for common sense in rural England in Chapters 3 and 5 below.

5 In summary, most anthropological writing on common sense has treated the concept theoretically, rather than ethnographically.

6 "I am suer Tyndale is not so farre besydis his comon sencis as to saye the dead bodye hereth cristis voyce" (Juhász 2014).

7 Although it is widely believed by the British public that this nomenclature indicates that the House of Commons is a deliberative chamber for common people (as opposed to the House of Lords, which is reserved for the nobility), in reality "Commons" descends from the word "communes" which in the thirteenth and fourteenth centuries was merely the word for an association or confederacy (Pollard 1920, 107–108).

8 *Chalk and cheese* refers to a fundamental and inimical difference. *To make a mountain out of a molehill* is to overreact to a slight issue.

9 The Terrae-filius was an officially-appointed satirist at the University of Oxford, appointed by the proctors, a position that existed from 1591 until 1763 (Dougill 1998, 273).

10 "The northern grave extended to the common of Euston" (Trans. Liam Lewis, personal communication).

11 "Certainly, all other sorts of gifts, such as land, tenements, pastures, or commons, or personal property …". An extract from one of Chaucer's tales, the full passage reading: "A wife is truly God's gift. Certainly, all other sorts of gifts, such as land, tenements, pastures, or commons, or personal property – all are gifts of Fortune, that fleet by as a shadow on a wall." (Trans. Lewis op. cit.).

12 "They put the land to common pasture [sharing it between them], that which lay nearest to hand …" (Trans. Lewis op. cit.).

13 McFate and Fondacaro, seeking to defend the Human Terrain Systems used by the Unites States armed forces, argue that:

> The military does not make foreign policy, but is constitutionally required to execute that policy. Providing them with the requisite knowledge to do so efficiently, carefully and with minimal loss of life is simple common sense. If Professor González or others have concrete, practical suggestions for other means to achieve this goal – bearing in mind that the military has no control over US foreign policy – we would certainly welcome this input.
>
> (2008, 27)

The pragmatic thrust of the phrase is very evident here.

14 There are a great many examples of this, and the scholars chosen above are selected merely for illustrative purposes. For example, Webb *et al.* define Bourdieuan heterodoxy thus: "The set of beliefs and values that challenge the status quo and *received wisdom – or common sense – within a particular field*" (2002, xiii), clearly intimating that the two are one and the same. Ann Stoler writes

> [attending to the unwritten] seeks to identify the pliable coordinates of what constituted colonial *common sense* in a changing imperial order in which social reform, questions of rights and representation, and liberal impulses and more explicit racisms played an increasing role. As imperial orders changed, so did common sense.
>
> (Stoler 2009, 3)

Imre Szeman claims: "Entrepreneurship exists in the twenty-first century as a commonsense way of navigating the inevitable, irreproachable, and apparently unchangeable reality of global capitalism" (Szeman 2015, 473).

15 The fact that Williams is using "culture" here – in the narrow sense of "high culture" – in quite a dissimilar sense to how anthropologists normally conceive of it, is for Crehan a crucial factor behind this misunderstanding of hegemony, as is how Williams' summary provides a clear and precise reading of what is – in Gramsci's original writings – a decidedly complex and difficult idea.

16 Folklore, for Gramsci, was essentially the popular counterpart to elite discourse – the largely fixed, but widely held traditions, beliefs, and worldview of common people (Gramsci 1985, 188–190, quoted in Crehan 2002, 106–107). It is, I suggest, very close to how Geertz, Beteille, Herzfeld and others describe common sense.

17 See the discussion of Paxman's quotation at the beginning of this chapter.

18 These "three orders" were the prevailing mode of social organisation across Christendom throughout the Middle Ages, finding clearest expression in the *Ancien Régime* of France (Duby 1982). Although they never prevented social mobility in Britain, the three orders nonetheless have a direct role on the political structure of the British State – with Parliament being divided into the Lords Spiritual (the clergy) and Temporal (the nobility), and the Commons.

19 The letters here refer to the stages of Wierzbicka's compound definition of common sense (see above).

References

Amherst, Nicholas. 2004. *Terrae-Filius, Or, The Secret History of the University of Oxford.* Edited by W. E. Rivers. Newark: University of Delaware Press.

Aristotle. 1957. *On the Soul.* Edited by W. H. Hett. Cambridge, MA; London: Harvard University Press.

Ashley Cooper, Anthony, Earl of Shaftesbury. 1709. *An Essay on the Freedom of Wit and Humour – a Letter to a Friend.* Edited by Jonathan Bennett. www.earlymoderntexts.com. www.earlymoderntexts.com/assets/pdfs/shaftesbury1709a_1.pdf.

Bayer, Thora Ilin. 2008. "Vico's Principle of Sensus Communis and Forensic Eloquence". *Chicago-Kent Law Review*, Symposium: Recalling Vico's Lament: The Role of Prudence and Rhetoric in Law and Legal Education, 83 (3): 1131–1155.

Beteille, A. 1996. "Sociology and Common Sense". *Economic and Political Weekly*, 35/37, 31 (35/37): 2361–2365.

Bloch, Maurice E. F. 1998. *How We Think They Think: Anthropological Approaches to Cognition, Memory, and Literacy.* 1st edition. Boulder, CO: Routledge.

Bogdan, R. J. 1991. *Mind and Common Sense: Philosophical Essays on Commonsense Psychology.* Cambridge: Cambridge University Press. http://ebooks.cambridge.org/ebook.jsf.

Bourdieu, Pierre. 1977. *Outline of a Theory of Practice.* Translated by Richard Nice. Cambridge Studies in Social and Cultural Anthropology. Cambridge University Press. www.cambridge.org/core/books/outline-of-a-theory-of-practice/193A11572779B478F5BAA3E3028827D8.

Bourdieu, Pierre. 2008. *The Bachelor's Ball: The Crisis of Peasant Society in Béarn.* Translated by Richard Nice. Cambridge: Polity.

Bugter, S. E. W. 1987. "Sensus Communis in the Works of M. Tullius Cicero". In *Common Sense: The Foundations for Social Science*, edited by F. van Holthoon and D. Olson. Lanham; New York, NY; London: University Press of America.

Burke, E. 2005. *Reflections on the Revolution in France.* London: Penguin Books.

Chisholm, R. M. 1982. *The Foundations of Knowledge.* Minneapolis: The University of Minnesota Press.

Comaroff, J., and J. L. Comaroff. 1991. "Of Revelation and Revolution", Vol. I. In *Christianity, Colonialism, and Consciousness in South Africa.* Chicago, IL: University of Chicago Press.

Crehan, K. 2002. *Gramsci, Culture, and Common Sense.* Berkeley: University of California Press.

Descartes, René. 1709. *Discourse on the Method of Rightly Conducting One's Reason and Seeking Truth in the Sciences.* Edited by Jonathan Bennett. www.earlymoderntexts.com. www.earlymoderntexts.com/assets/pdfs/shaftesbury1709a_1.pdf.

Descola, Philippe, and Gisli Palsson, eds. 1996. *Nature and Society: Anthropological Perspectives.* 1st edition. London: Routledge.

Dougill, John. 1998. *Oxford in English Literature: The Making, and Undoing, of "the English Athens".* Ann Arbor: University of Michigan Press.

Duby, Georges. 1982. *The Three Orders: Feudal Society Imagined.* Chicago: University of Chicago Press.

Easthope, Antony. 2004. *Englishness and National Culture.* London: Routledge.

Elio, R. 2002. *Common Sense, Reasoning, and Rationality.* Oxford: Oxford University Press. www.oxfordscholarship.com/view/10.1093/0195147669.001.0001/acprof-9780195147667.

Feierman, S. 1990. *Peasant Intellectuals: Anthropology and History in Tanzania.* Madison, WI: The University of Wisconsin Press.

Ferguson, Robert A. 2000. "The Commonalities of Common Sense". *The William and Mary Quarterly* 57 (3): 465–504. https://doi.org/10.2307/2674263.

Fox, K. 2005. *Watching the English: The Hidden Rules of English Behaviour.* London: Hodder & Stoughton.

Fuentes, A. 2012. "Anthropology and the Assault on Common Sense: Critical Thinking About Being Human Is a Useful Hobby". In *The Huffington Post – The Blog* (blog). 2012. www.huffingtonpost.com/american-anthropological-association/anthropology-and-the-assa_b_1834358.html.

Geertz, Clifford. 1983. *Local Knowledge.* New York, NY: Basic Books.

Gramsci, A. 1971. *Selections from the Prison Notebooks.* London: Lawrence & Wishart.

Gramsci, A. 1985. *Antonio Gramsci: Selections From Cultural Writings.* Edited by D. Forgacs and G. Nowell-Smith. London: Lawrence & Wishart.

Gregoric, P. 2007. *Aristotle on the Common Sense.* Oxford: Oxford University Press. www.oxfordscholarship.com/view/.

Herzfeld, Michael. 2001. "Common Sense, Anthropology Of". In *International Encyclopedia of the Social & Behavioural Sciences*, 2283–2286. Science Direct.

Howes, David. 2003. *Sensual Relations: Engaging the Senses in Culture and Social Theory/ David Howes.* Ann Arbor, MI: University of Michigan Press.

"Italian Translation of 'Common Sense'". 2017. In *Collins English–Italian Dictionary.* Scotland: HarperCollins. www.collinsdictionary.com/us/dictionary/english-italian/common-sense.

Johnson, B. 2013. *The Role Ethics of Epictetus: Stoicism in Ordinary Life*. Lanham: Lexington Books.

Johnston, George Alexander, Thomas Reid, Adam Ferguson, James Beattie, and Dugald Stewart. 1915. *Selections from the Scottish Philosophy of Common Sense,* edited, with an Introduction, by G. A. Johnston. Chicago, London: The Open Court Publishing Company, 1915.

Jolly, R. 2000. *Jackspeak: An Illustrated Guide to the Slang and Usage of the Royal Navy and Royal Marines Including the Submarine Service and the Fleet Air Arm*. Great Britain: FoSAMA.

Juhász, G. M. 2014. *Translating Resurrection: The Debate between William Tyndale and George Joye in Its Historical and Theological Context*. Netherlands: Brill.

Kant, I. 1914. *Kant's Critique of Judgement*. Translated by J. H. Bernard. 2nd edition. London: Macmillan.

Kramnick, Isaac, and Michael Foot. 1987. "Editors' Introduction: The Life, Ideology and Legacy of Thomas Paine". In *Thomas Paine Reader,* edited by Isaac Kramnick and Michael Foot, Reprint edition. Harmondsworth, MDX; New York, NY: Penguin Classics.

Kuklick, B. 2012. "Principal Events in Paine's Life". In *Paine – Political Writings,* edited by B. Kucklick. Cambridge: Cambridge University Press.

Kurtz, D. 2013. "Gramsci, Antonio". In *Theory in Social and Cultural Anthropology: An Encyclopedia,* edited by R. J. McGee and R. L. Warms. Los Angeles, London, New Delhi, Singapore, Washington DC: Sage Reference.

Lemos, N. 2009. *Common Sense: A Contemporary Defense*. Cambridge: Cambridge University Press. http://ebooks.cambridge.org/ebook.jsf.

Marx, Karl, and Friedrich Engels. 2000. "The German Ideology". In *Karl Marx: Selected Writings,* edited by D. McLellan. 2nd edition. Oxford: Clarendon Press.

McFate, M., and S. Fondacaro. 2008. "Cultural Knowledge and Common Sense in Anthropology". *Today* 24 (1): 27.

Nadel, S. F. 1951. *The Foundations of Social Anthropology*. London: Routledge.

Paine, Thomas. 2012a. "Common Sense". In *Political Writings – Cambridge Texts in the History of Political Thought,* edited by B. Kuklick. Cambridge: Cambridge University Press.

Paine, Thomas. 2012b. "Rights of Man, Part I". In *Political Writings – Cambridge Texts in the History of Political Thought,* edited by B. Kuklick. Cambridge: Cambridge University Press.

Paine, Thomas. 2012c. "Rights of Man, Part II". In *Paine: Political Writings – Cambridge Texts in the History of Political Thought,* edited by B. Kuklick. Cambridge: Cambridge University Press.

Parkin, Robert. 2009. *Louis Dumont and Hierarchical Opposition*. Oxford: Berghahn Books.

Paxman, J. 2001. *The English: A Portrait of a People*. London: Penguin.

Petty, Karis. 2017. "Sensing the South Downs." *South Downs National Park* (blog). 4 July 2017. www.southdowns.gov.uk/how-the-visually-impaired-perceive-wild-places/.

Pierce, Charles. 1905. "What Pragmatism Is". *The Monist* 15 (2): 1–21.

Pollard, A. F. 1920. *The Evolution of Parliament*. London: Longmans.

Pollock, J. 2002. "The Logical Foundations of Means-End Reasoning". In *Common Sense, Reasoning, and Rationality,* edited by R. Elio. Oxford: Oxford University Press.

Porter, Roy. 1992. *The Scientific Revolution in National Context*. Cambridge: Cambridge University Press.
Proctor, L. 2012. *Anthropology and Common Sense: Fuentes, Béteille, and Public Anthropology*. http://anthrocharya.com/2012/09/01/common-sense/.
Reid, T. 2005. *An Inquiry into the Human Mind*. Edited by J. Bennett. www.early moderntexts.com. www.earlymoderntexts.com.
Ryan, A. 2012. *The Making of Modern Liberalism*. Princeton and Oxford: Princeton University Press.
Scott, James C. 1985. *Weapons of the Weak: Everyday Forms of Peasant Resistance*. New Haven, CT; London: Yale University Press. http://quod.lib.umich.edu/cgi/t/text/pageviewer-idx?c=acls;cc=acls;rgn=full%20text;idno=heb02471.0001.001;didno=heb02471.0001.001;view=image;seq=3;node=heb02471.0001.001%3A1;page=root;size=100.
Silverman, M. 2000. "Custom, Courts, and Class Formation: Constructing the Hegemonic Process Through the Petty Sessions of a Southeastern Irish Parish, 1828–1884". *American Ethnologist* 27 (2): 400–430.
Simms, Brendan. 2008. *Three Victories and a Defeat: The Rise and Fall of the First British Empire, 1714–1783*. Harmondsworth: Penguin.
Simpson, J. A., and E. S. C. Weiner, eds. 2009. "Oxford English Dictionary". 2nd edition. Oxford: Clarendon Press.
Sober, E. 2008. "Empiricism". In *The Routledge Companion to Philosophy of Science*, edited by S. Psillos and M. Curd, 129–138. London; New York, NY: Routledge.
Stoler, A. 2009. *Along the Archival Grain: Epistemic Anxieties and Colonial Common Sense*. Princeton: Princeton University Press.
Strang, Veronica. 2005. "Common Senses: Water, Sensory Experience and the Generation of Meaning". *Journal of Material Culture* 10 (1): 92–120. https://doi.org/10.1177/1359183505050096.
Szeman, I. 2015. "Entrepreneurship as the New Common Sense". *South Atlantic Quarterly* 114.3: 471–490. http://saq.dukejournals.org/content/114/3/471.full.pdf.
van Kessel, P. 1987. "Common Sense between Bacon and Vico: Scepticism in England and Italy". In *Common Sense: The Foundations for Social Science*, edited by F. van Holthoon and D. Olson. Lanham; New York, NY; London: University Press of America.
Vico, Giambattista. 1948. *The New Science of Giambattista Vico*. Translated by Thomas Goddard Bergin and Max Harold Fisch. Ithaca, NY: Cornell University Press. https://archive.org/stream/newscienceofgiam030174mbp/newscienceofgiam030174mbp_djvu.txt.
Watts, D. 2011. *Everything Is Obvious Once You Know the Answer*. London: Atlantic Books.
Webb, Jen, Tony Schirato, and Geoff Danaher. 2002. *Understanding Bourdieu*. London; Thousand Oaks, CA: Sage Publications Ltd.
Wierzbicka, A. 2009. *Experience, Evidence, and Sense: The Hidden Cultural Legacy of English*. Oxford, New York: Oxford University Press. www.oxfordscholarship.com/view/.
Williams, Raymond. 1977. *Marxism and Literature*. Marxist Introductions. Oxford; New York, NY: Oxford University Press.
Wilson, C. 2012. *Descartes' Meditations: An Introduction*. Cambridge: Cambridge University Press.

2 What is common sense?

Common sense rests upon a profound tension between academic and popular knowledge in English society. While scholars inspired by classical precedent have used the term to reflect upon universal human capacities for reason and morality, the vernacular use of the term is oriented around pragmatic, situated understanding that solves problems and scorns such lofty abstractions. The etymology of this phrase, and the work of other social scientists including anthropologists and linguists point to a need for common sense to be examined ethnographically – that is, through sustained observation of its use in ordinary, day-to-day surroundings – if a sufficient understanding of its power and meaning is to be achieved.

Below, I examine four instances where common sense became visible in the course of my fieldwork – in one specific landscape: the Mid Yare Valley and its watershed. The Yare is one of the principal rivers of the Broads, and the middle of its course – between the suburbs of Norwich and the wide grazing marshes of Halvergate – is carefully maintained by a wide range of different land managers. As discussed in my introduction, the social landscape of the Broads is highly fragmented, so it was not possible to construct the sort of holistic account so characteristic of classic ethnographies without distorting my own material considerably. Instead, what I produce here is more of a *montage* (Grimshaw 2001, 11) – a set of discontinuous snapshots of common sense in action. Each illustrates how common sense is deployed in distinct social situations, underscoring its pragmatic, materially situated quality.

Common sense as a vernacular object

I arrived at the shooting estate[1] early on a bright and crisp winter morning. Walking boldly up the crunchy gravel driveway, I was somewhat intimidated by the sight of the place. A large, well-maintained manor amidst carefully tended gardens, this was one of the most formidable estates I had yet visited. I had been told to come to the back door, so I rounded the side of the building, and I saw the estate's marshes stretching out in front of me for the first time – flocks of unidentified birds wheeled in the distance, and the whole

Yare Valley was wreathed in mist. I passed a brace of pheasants hanging on a hook and a pile of boots by the door, and knocked.

Despite its grand and traditional exterior, the manor was comfortable and modern inside. The owner of the estate, C, was a straight-talking and welcoming chap in his 50s, with a booming voice but an affable manner, who briskly led me through to the kitchen and made me a coffee. As we sat waiting for the estate's Conservation Officer, C talked in glowing terms about all the raptors he'd encouraged onto the Estate – gesturing to some feeding stations he'd set out where the smooth lawn of the manor's gardens met the rough tussocks of the marsh. When the Conservation Officer arrived, he was given a hot mug of tea, and we moved through to one of the sitting rooms and sat down at a low table. I had a view into the main hall, whose walls were covered with the mounted heads of African wild game.

I interviewed C and the Conservation Officer about the management of the estate, which prompted C to discuss how there was great difficulty in spreading knowledge of how to farm with wildlife in mind, even within the industry. "Farming is not rocket science," he said, "but we currently don't have that centre of expertise." He then said that he felt some sort of central "think tank" of sensible, knowledgeable farmers who could provide practical advice would be a great way to fill this knowledge gap. This was contrasted to what was dubbed "hedgehog science" – whereby scientific experimentation was used to prove what was already sensible – "a matter of common sense" – for practically experienced landowners. When I asked for an example, the owner mentioned a recent experiment at a local nature reserve regarding pest control – that found shooting foxes caused a steep recovery in the success of breeding birds. This was the obvious outcome for both C and the Conservation Officer at the manor – "Any gamekeeper could tell you that!" – but a senior official at the nature reserve had exclaimed, "We've now got the science to prove it!" This was felt by C to be utterly unnecessary. Why spend time and money trying to prove that something worked, when it was already common sense for anyone who had the requisite experience that it did?

What we see in this case is common sense being paired with an antithesis: science. As a form of knowledge, common sense is everything science is not: intuitive, discretionary, vernacular, pragmatic. While both can serve pragmatic interests, science requires dedicated experimentation on point of principle – because it is *empirical* – while common sense merely arises organically from personal experience of what works. "Hedgehog Science" has such faith in its own method that it does not consider any other technique for gaining insight to have sufficient power to prove anything – so even when common sense indicates that shooting foxes will lead to an increase in bird numbers, scientists will nonetheless insist upon testing the hypothesis. Such a duplication of effort on point of epistemic principle shows a lack of common sense. The common sense approach would be, by contrast, simply to take on board what is already known. *If it ain't broke* – so the saying goes – *don't fix it.*

This incident also reveals how the tensions described earlier in this chapter, between intellectuals and vernacular voices, play out in the field. Here, the disagreement was between common-sense led farmers and gamekeepers, and science-led statutory and conservation bodies, but it could equally have been ordinary local residents versus university-educated bureaucrats, or similar – this oppositional arrangement was a common trope in the Broads. It articulated the tensions between different professional fields over how the countryside should be managed, each of which has their own distinct class status in English society. University-educated professionals are usually considered middle or upper-middle class, while gamekeepers are usually working class. Farmers and landowners – despite being from traditional upper-class families – would demonstrate far greater solidarity and sympathy with the common-sense view of their working class staff, than they would with their middle-class interlocutors in conservation charities or government. This demonstrates the provinciality and specificity of common sense; it is by no means a universal attribute of human reason, but rather a specific trait that is exhibited by particular people, in particular places, doing particular things. This becomes even clearer in the following excerpt.

⋆ ⋆ ⋆

Away from the lakes and fens of its wide, slow rivers, East Norfolk is largely agricultural. Countless acres of land are given over to the growing of sugar beet, wheat, herbs, and root vegetables. When travelling on the main roads that criss-cross the county, it was common to be caught behind large agricultural vehicles – trucks, tractors, combine harvesters – travelling from farm to farm, or taking produce to be processed or sold, especially during the sowing and harvesting season. The timetable of the X2 bus route between the regional capital of Norwich and the idyllic little market town of Beccles – where I spent much of my time in early 2015 – was therefore only ever approximate.

The movement of the agricultural vehicles, and the changes in the land they were tending, had remained something of a closed book to me until the spring of that year. I had spent many months attempting to speak to farmers, and yet I found it very difficult to secure an "in" – most were unwilling to speak to me, and those who did could only spare a couple of hours of their time. As such, the fields and the movement of equipment between them were a text I could not read; having not been to agricultural college, nor born into a farming family, Norfolk's agriculture was much like my view of it through the windows of the X2 – distant, indistinct, a pleasant view to be quietly enjoyed in passing.

I met P at the 2015 Norfolk Farming Conference; he questioned MP Liz Truss on her stance regarding the influence of agribusiness and supermarkets on farming, and I decided I had to speak to him. P was the first farmer I met who was happy to talk, and he had a lot to say about the state of British agriculture. But first and foremost, he offered to give me a tour of the fields of

Norfolk, to – in his words – help me "get my eye in" – to be able to recognise agricultural processes.

P took great pleasure in helping me to see the landscape the way he did. He pointed out how the different shades of green in wheat and barley fields indicated if they needed an application of fertiliser; leaves that had a yellow tint – looking "a bit *starey*" – would turn a dark green with an application of nitrogen. When we saw a suckling herd, P explained the impact of stocking levels on whether you could keep cattle outdoors over winter. As it was spring, the root crop fields were all ploughed and ready to be sown – P pointed out the distinctive bed structure for different crops, including carrots (three raised beds, with one deep trench), potatoes (one bed with two lines of raised earth by a deep trench), and sugar beet (lines traced in the soil). Great sheets of clear plastic also covered the beds. P described these as miniature greenhouses, warming the soil and bringing the crops on earlier.

P and I later moved on to discuss socio-economic trends in agriculture. Speaking about the shortcomings of existing sustainable food-growing projects, P mentioned how he'd been invited to join a suburban grower's co-op as an expert member of the team. In P's view, although this group was "aspirational", they were not prepared to actually learn how to cultivate properly for sale, and so their business was not horticulturally sustainable – being merely "gardening on a grand scale". Projects of this kind, P felt, simply scaled up techniques of food growing for domestic use, in order to increase production. "If you're running a garden, it's fine, but if you use those techniques on a field, it's backbreaking. If you use machines, it's still hard, but it's not as hard." P eventually left the co-op; he had decided to plough up the field at the beginning of the year, something he felt was "simple common sense", but the organisers had felt they should have been consulted before this decision was taken. This reaction had been, for P, the final straw.

Here, we see the *specificity* of common sense. In referring to a particular task as "simple common sense", P is presenting it as routine, obvious, and necessary. No comment or discussion was required, in his view – just like one wouldn't need to stipulate that someone shouldn't light a bonfire under a tree with low-hanging branches, or leave the handbrake off on a car parked on a slope. His colleagues, however, disagreed. They wished to be consulted. P saw this as patently ridiculous: evidence of their lack of engagement in the practice of horticulture and the common sense of farming.

Common sense, here, has clear limits – it leaves out certain people, and is tied to others (see Figure 2.1). P was not suggesting that his interlocutors lacked all capacity for reason; he was rather suggesting that they were not as familiar with how horticulture takes place, and therefore lacked the sense common to it. This would be bad enough, but they also lacked awareness of their own inexperience. By attempting to micro-manage P from a position of ignorance, they demonstrated that they did not have the flexible, practical mentality required to deal properly with problems in a

Figure 2.1 A broad taskscape. Common sense arises from particular landscapes, and implicates specific people, while excluding others. As we shall see in subsequent chapters, this pattern is inflected within property rights – from private ownership, to rights of navigation along major rivers.

common-sensical fashion – in short, they breached certain vital norms. As such, P decided that further collaboration would be a waste of time, and ended his involvement.

* * *

Whitlingham Country Park is Norwich's gateway to the Broads – 35 hectares of woodland, meadows, and open water, lying south-east of the city. Whitlingham Broad, despite falling within the National Park, is not actually a "proper" broad, as it is the flooded remnants of a modern gravel pit, rather than a medieval turbary. Despite being geologically atypical, it is nonetheless an important site for the work of the Broads Authority; who co-manage the site with the Whitlingham Charitable Trust. The large visitor centre is generously supplied with leaflets and posters advertising the attractions in the wetlands beyond; the network of paths that spread out across the Park is lined with information boards that detail the natural history of the place.

Wardens employed by the Broads Authority tend the site, and from here they patrol the rest of the Mid Yare Valley. I was fortunate enough to be allowed to shadow F, one of the most experienced wardens on the Broads Authority's staff. He took me on a tour of the Park, as he completed his usual rounds, inspecting the paths and checking water levels, repairing fences and speaking with the local boaters. Towards the end of the day, we were approached by a man who lived on a houseboat moored up on the riverbank between the Yare and the broad. Clearly distressed, he complained that he'd been subjected to verbal abuse by another boater. F discerned that the second boater had been speeding; the houseboater had called him to slow down, leading to a hostile verbal exchange by the two men.

F headed out to confront the man in question, and took me along so that I might see what he felt to be an important part of his work: keeping the peace on the river. The Broads Authority has a statutory responsibility for navigation on the Broads; so their wardens are responsible for enforcing a wide range of bylaws concerning boat traffic upon the water. Conservation occupies only a third of a warden's time – the remaining two thirds is taken up with overseeing navigation on their beat. We drove up to where the houseboater had told us the other man had moored up. F got out, and asked me to stay in the van. When he returned, having warned the man concerned, F started to explain what he felt was crucial to keeping the peace effectively. The Broads Authority only has a limited budget and amount of time to prosecute people who break their regulations. As such, it falls upon wardens such as F to exercise their discretion – to "use their common sense" – in deciding whether there is sufficient evidence and public benefit to taking allegations of wrongdoing further. The method F had developed to assist with this was what he called "the Attitude Test". If he spoke to someone who had flouted a regulation, and they gave evidence of a genuinely positive attitude – i.e. they were polite, showed remorse, and made commitments not to break the same rule again – he would let them off. But repeat offenders, or people who seemed to have a poor attitude – who were rude, or antagonistic, or refused to apologise – would have charges brought against them. "You've just got to hope, when you're dealing with people every day, that they will just show a bit of common sense, and work with you."

F's approach to his responsibilities fits cleanly with Wierzbicka's definition of common sense. F was faced with an array of problems – a limited amount of resources to support legal action; a responsibility for keeping the peace; a man speeding and verbally abusing another river user – and took the approach of making judgement calls in a very short space of time to ensure the best possible outcome. All of this pertains to a common resource (namely, the River Yare). Such an approach would be impossible, if F himself lacked common sense.

But equally, it is arguable that the people to whom the Attitude Test is applied *also* need common sense. When presented with a problem – being confronted by a Broads Authority warden over a misdemeanour – in a short space of time they need to decide how to respond. Do they behave in a conciliatory, respectful fashion, or do they act aggressively? The penalty for behaving aggressively is obvious: information regarding the Broads Authority's statutes and penalties is circulated to all boat owners and renters on the Broads. If they make an incorrect choice – as deemed by the warden – then it is their attitude towards the river, and the warden, that is at fault. As such, F's use of the Attitude Test doesn't just evince his own, common-sense approach to solving problems, but also the importance of common sense as an enforceable norm in the Broads.

* * *

It was the last day of my fieldwork. RSPB Strumpshaw Fen – a nature reserve where I spent four months as a residential volunteer[2] – was holding a barbeque in honour of its 30 volunteers, who assisted the permanent staff with visitor engagement and the management of the reserve. People were crowding around the grills, or standing a little way off, talking to one another while sitting on plastic garden chairs. The air was convivial, but the scrupulously private approach everyone took to the food drew my attention. Everyone brought, cooked, and ate, their own food. Although there was a sense to it – not everyone wanted to eat the same things, and this approach allowed for more personal choice – I was struck by how this contrasted markedly with the communal eating seen in other ethnographic cases (e.g. Carsten 1995). In that moment I realised what concept brought all my experiences in Norfolk together: *common sense*.

My boss on the reserve, O, was standing talking to my parents, who'd arrived to help me move my possessions out of the volunteers' cottage. After I cooked the salmon we were having, I went over to join them, and told them about the angle I'd decided to take with my thesis. O smiled, and looked into the middle distance. "There's only two things I look for in residential volunteers," he said, "enthusiasm and initiative. I can teach them everything else." At first, this response might seem opaque – what relationship do enthusiasm and initiative have with common sense? It was clear from the way that O spoke, he wasn't disagreeing or correcting me: he offered his words as a meditation on mine. So what did O mean?

If we consider O's remarks in the context of the previous cases – F's seeking of the right attitude in instituting shared norms, C's emphasis upon direct, practical experience of the landscape, and P's experiences with those who lacked common sense, a possible explanation emerges. The two things O chose as important prerequisites for being a good residential volunteer – *enthusiasm* and *initiative* – are both prerequisites for the acquisition of the reserve's common sense. Enthusiasm is a particular kind of attitude, a set of internal dispositions consistent over time. Just as F sought out general cooperativeness as the right sort of attitude on the River Yare,[3] enthusiasm was normative on the reserve. The hours are long, the work is physically demanding and occasionally dangerous, while the landscape itself can be very unforgiving. Enthusiasm for the work of the reserve was therefore crucial, both as a personal motive for individual staff members, and as an important source of bonding within the team. Individuals who did not share that enthusiastic attitude would thus be alienated, both from the work and the other workers, and would struggle to work on the reserve effectively – and thus could not acquire the sense common to those who worked there.

If enthusiasm speaks of a normative attitude, initiative speaks to pragmatism. Initiative – often used in conjunction with the phrase "self-starter" – was highly prized on the reserve, and anyone exhibiting it was praised. It could manifest in something as simple as passing a colleague working on building a pond-dipping platform[4] screws as and when he required them, without being asked (see Figure 2.2). If you could correctly anticipate the needs of one's colleague, and assist his work, then you were showing initiative. In short, initiative is being able to take *prompt, practical action independently* (Simpson and Weiner 2009) – something that was indicated when another warden on the reserve, during an interview, described common sense as "doing what is required without thinking or being taught".[5] What is

Figure 2.2 Working together. Knowing when to pass a friend a tool as they needed it, or doing an obvious task that needed doing without being asked, were all signs of "common sense".

striking about P's story above is that when he attempted to use his initiative – by ploughing up the field – he was castigated by his colleagues; he responded by saying that this indicated that they lacked common sense.

O's confidence that, aside from these two things, he could teach *anything else*, is also suggestive. Common sense is not just a form of knowledge that is non-academic, but – and this was confirmed by numerous informants – it is thought to be something that arises organically, without needing to be formally taught. When C discussed his idea for a centre for expertise in farming, he was not suggesting an institute for teaching common sense to his peers; good farming, in C's view, was not a wholly separate field to academic knowledge per se.[6] Rather, what distinguishes landowners, farmers, and gamekeepers from the practitioners of "hedgehog science", was the pragmatic sensibleness of the former, in contrast to the excessive empiricism of the latter. Scientific knowledge, farming skills, and conservation techniques could all be taught – what could not be taught was the kind of pragmatic attitude that allowed you to work with your peers, and develop the right *nous*[7] to sense what actions were right and wrong, without being actively told. All these ethnographic cases indicate a key context for the exercise of this colloquial understanding of common sense: namely a *shared working environment* – a landscape shaped by working together. The role of common sense in shaping – and failing to shape – common norms will be examined in greater detail in Chapter 5, while the assumed unteachability of common sense will be considered in Chapter 4.

Working in a common environment has long been a popular topic of English nature-writing (Leighton 1992; Collis 2009; Macfarlane 2015), as well as anthropological enquiry (e.g. Malinowski 1935; Evans-Pritchard 1969). But the most relevant scholarly oeuvre for thinking through the issues raised above – time, landscape, practical engagement, and phenomenal experience – is that of Tim Ingold. In his seminal text *The Perception of the Environment* (2000), Ingold develops a thoroughgoing critique of what he deems to be characteristically "Western" styles of thought and action – focused upon the "mastery" of the natural world (ibid., 321) that is predicated on a fundamental Cartesian divide between the mind and the body, the human and the natural (Ingold 2017a, 4, 2000, 260). In its place, Ingold draws on the philosophical traditions of hunter-gatherer societies (such as the Mbuti, the Cree, and others) and the Gibsonian notion of "affordances" (Gibson 1977, 2000, 3, 5) to suggest that "we" Westerners should reject the nature culture divide, and that we should "follow the lead of hunter-gatherers in taking the human condition to be that of a being immersed from the start, like other creatures, in an active, practical and perceptual engagement with constituents of a dwelt-in world" (Ingold 2000, 42). Ingold's phenomenological approach to practical activities and materiality has informed a broad swathe of subsequent scholarship in both anthropology and archaeology (e.g. Tilley 2004; Henare *et al.* 2006; Berkes 2012; Vergunst 2012; Vergunst and Vermehren 2012). Ingold develops the *taskscape* as the precise correlate to the word *landscape* – "Just as the landscape is an array of related features, so – by analogy – the taskscape is an array of related activities" (Ingold 2000, 195). The taskscape is the complete array of tasks – acts that are constitutive of dwelling – that go into a landscape. The taskscape is to labour, therefore, what landscape is to land. This connection of landscape to taskscape and labour therefore entails a consideration of time – leading Ingold to conclude that each momentary apprehension of the taskscape in the present reveals vistas onto the past and future. The taskscape is, in Ingold's terminology, a "plenum" – it has no holes or gaps (Ingold 2000, 191).

Ingold's work in *Perception* helps us to understand the cultural underpinnings to common sense in Broadland, by raising a series of important questions. First, his characterisation of a dwelling perspective of hunter-gatherer societies as a counter to the prevailing perceptual schemes in the "Western" world invites us to consider: How do the kinds of common sense we find among land managers in Norfolk relate to the "Western" sensibilities that Ingold critiques? Second, his characterisation of the taskscape – the sum of actions that are constitutive of dwelling in a place – encourages us to reflect upon the totality of practices that contribute to the socio-ecology of Broadland. Finally, his affirmation that the taskscape is a plenum invites further questions: What is the role of conflict, tension, power, and change within the taskscape of the Broads? How do labour, work and experience of the environment relate to processes of economic and political change?

In many respects, vernacular common sense differs from the "Western" outlook identified by Ingold – it presumes precisely the kind of "active, practical, and perceptual engagement" with natures' "affordances" that Ingold so vividly describes in non-Western contexts. The shared working environment that common sense requires correlates precisely with the "taskscape" identified by Ingold; we shall examine the Broadland taskscape in greater detail below.

But where common sense and the Ingoldian "dwelling perspective" differs is with respect to economic and political change. The work of Anna Tsing indicates the importance of attending to sites of friction between different forms of knowledge and ways of living – as she remarks:

> It would be easier for everyone if rational deliberations always converged in common understandings. But even those of us who believe that some knowledge claims are better than others have difficulty in denying that even the best ones retain a certain incommensurability. This is because knowledge claims emerge in relation to concrete problems and possibilities for dialogue – productive features of friction.
>
> (Tsing 2005, 10)

As we shall see, academic accounts of the Broads depict it as a site of "ruination" – a peasant landscape disrupted and transformed by capitalist modes of production and urbanism, much like the *matsutake*-growing forests of Japan

(Tsing 2015, 180–187, 189–190). By examining common sense ethnographically, we discover it is used to imagine a landscape that is not so much a plenum, but rather a sort of "landscape of unintentional design" (ibid., 286), where the lack of central orchestration and the presence of disruptive political forces indicate the potential for gaps and holes within which useful products – and dynamic problem-solving – emerge. As we shall see, despite land managers' sensing the marshy surface of Broadland in a highly kinaesthetic, "haptic" way (Ingold 2017b), movement and management through field and fen is not free of friction, either historically or in the present day.

Common sense in vernacular use

Today, among the inhabitants of the Broads, common sense is used to talk about a particular kind of pragmatism, an experiential attitude geared towards problem-solving through trial and error, whose precise limits, values, and associations will be traced further in the following chapters. What became apparent as I worked in the fields and fens of the Broads, was that the people I worked alongside experienced common sense quite differently to the long tradition of academic thought detailed in the previous chapter, while nonetheless owing a clear debt to it. Farmers, conservationists, gardeners, and gamekeepers were all – through the practical demands of working in specific surroundings – in dialogue with those surrounds, as well as with the academic philosophising they sought to eschew. Their shared experiences of all kinds accreted into common senses particular to their own day-to-day working lives. These ethnographic cases amply illustrate the central role of labouring in particular places, with particular people, plants, animals, objects, and materials, in shaping the vernacular conception of common sense. Invocations of common sense by land managers clearly chime with Ingold's exhortation to embrace actual, perceptual engagement with a shared taskscape, even if that taskscape is more haphazard, tense and gap-riddled, than a holistic plenum would be.

With this in mind, we arrive at the following – ethnographically particular – definition of common sense, as a complement to those already explored in the previous chapter:

> Common sense is a form of ***pragmatic attitude***, developed through daily interactions with the shared material and social context or "taskscape" – that is, the ***commons*** – of a defined group of people.

This is a working definition in two senses: it is developed for a specific function (to aid the understanding of social life in the Broads National Park), but it is also a definition developed and established through work in the common landscape I will now move to describe.

Notes

1 Most identities have been disguised with letters in this thesis.
2 The RSPB maintains properties on some of its larger reserves that allow for small numbers of full-time volunteers to live on-site for extended periods.
3 This recalls aspects of Wierzbicka's model above:

> l. "it is good if someone thinks like this
> m. all people can think like this
> n. it is good if people do things because they think like this
> o. it is bad if someone doesn't think like this."
>
> (Wierzbicka 2009, 10:351, original emphasis)

4 Pond-dipping is a naturalist pastime, whereby a net is used to fish small animals – larvae, tadpoles, snails, and so on – out of the pond in order that they might be studied. A number of wooden platforms had been built beside ponds to facilitate this popular activity.
5 Recalling other aspects of Wierzbicka's model:

> g. "I don't want something bad to happen
> h. because of this, I want to think about it for a short time
> i. if I think about it for a short time, I can know what I can do"
> j. when this someone thinks like this, after a very short time this someone can know it
> k. ***(because of this, this someone does something)***
>
> (Wierzbicka 2009, 10:351, original emphasis)

6 C had an agronomist, for example, who he employed to advise him on what crop to plant, etc.
7 Nous, prounced /naʊs/, is an informal English synonym for common sense, or practical understanding. Because it is such a close synonym, I have avoided using it in my definition here.

References

Berkes, F. 2012. *Sacred Ecology*. London: Routledge.

Carsten, J. 1995. "The Substance of Kinship and the Heat of the Hearth: Feeding, Personhood, and Relatedness among Malays in Pulau Langkawi". *American Ethnologist* 22 (2): 223–241.

Collis, John Stewart. 2009. *The Worm Forgives the Plough*. Edited by Robert Macfarlane. London: Vintage Classics.

Evans-Pritchard, Edward Evan. 1969. *The Nuer : A Description of the Modes of Livelihood and Political Institutions of a Nilotic People*. Oxford: Oxford University Press.

Gibson, James J. 1977. "The Theory of Affordances". In *Perceiving, Acting and Knowing: Toward an Ecological Psychology*, edited by Robert E. Shaw, John D. Bransford, and University of Minnesota, 67–82. Hillsdale, N.J: Erlbaum. http://readinglists.exeter.ac.uk/ssis/Anthropology/ANT1008-SOC1008/ANTSOC1008_5_cv.pdf.

Grimshaw, Anna. 2001. *The Ethnographer's Eye: Ways of Seeing in Modern Anthropology*. Cambridge: Cambridge University Press.

Henare, A., M. Holbraad, and S. Wastell, eds. 2006. *Thinking Through Things: Theorising Artefacts Ethnographically*. London: Routledge.

Ingold, Tim. 2000. *The Perception of the Environment: Essays on Livelihood, Dwelling and Skill*. London: Psychology Press.

Ingold, Tim. 2017a. "Five Questions of Skill". *Cultural Geographies*, April, 14744 7401770251. https://doi.org/10.1177/1474474017702514.

Ingold, Tim. 2017b. "Surface Visions". *Theory, Culture & Society* 34 (7–8): 99–108. https://doi.org/10.1177/0263276417730601.

Leighton, Clare. 1992. *The Farmer's Year*. Edited by Pat Jaffe. London: The Sumach Press.

Macfarlane, Robert. 2015. *Landmarks*. 1st edition. London: Hamish Hamilton.

Malinowski, B. 1935. *Coral Gardens and Their Magic: A Study of the Methods of Tilling the Soil and of Agricultural Rites in the Trobriand Islands*. Vols I and II. London: Routledge.

Simpson, J. A., and E. S. C. Weiner, eds. 2009. "Oxford English Dictionary". 2nd edition. Oxford: Clarendon Press.

Tilley, C. 2004. *The Materiality of Stone: Explorations in Landscape Phenomenology*. Oxford, UK: Berg Publishers.

Tsing, Anna. 2005. *Friction: An Ethnography of Global Connection*. United Kingdom: Princeton University Press.

Tsing, Anna. 2015. *The Mushroom at the End of the World: On the Possibility of Life in Capitalist Ruins*. Princeton: Princeton University Press.

Vergunst, J. 2012. "Farming and the Nature of Landscape: Stasis and Movement in a Regional Landscape Tradition". *Landscape Research* 37 (2): 173–190.

Vergunst, J., and A. Vermehren. 2012. "The Art of Slow Sociality: Movement, Aesthetics and Shared Understanding". *Cambridge Anthropology* 30 (1): 127–142.

Wierzbicka, A. 2009. *Experience, Evidence, and Sense: The Hidden Cultural Legacy of English*. Oxford; New York, NY: Oxford University Press. www.oxfordscholarship.com/view/.

3 Where is common sense to be found?

During the winter I spent in Norwich, I rented a room in a terraced house on Kett's Hill – a steep road close to the railway station, overlooking the River Wensum. Kett's Hill was named after the sixteenth-century farmer Robert Kett, who led a rebellion in 1549 against the enclosure of Norfolk's fields and rivers. Kett and his rebels captured Norwich, although they were routed by the Earl of Warwick and his mercenaries soon afterwards (Chandler 2012). Before his ill-fated rebellion, Kett had presented a petition to Edward VI, stating that: "We pray that Ryvers may be ffree and common to all men for ffyshyng and passage …" and, perhaps indicating the involvement of the Broadlanders in the rebellion, that "redegrounde and meadow-grounde may be at such price as they were in the first yere of Kyng Henry the VII" (Ewans 1992, 12). Such demands indicate: it was the enclosure of common resources along Broadland's rivers – fisheries, navigation rights, reedbeds, and meadows – that prompted Kett and his fellows to rise up against the establishment. I didn't realise that one of my future colleagues at RSPB Strumpshaw Fen rented a place just across the road from me. He'd later confess that he'd never buy a property there; the hill was pockmarked with sinkholes, and there was a significant risk of subsidence. The sinkholes, and Robert Kett's doomed fight against enclosure, came to mirror one another in my mind. Both were integral parts of local landscape, hidden beneath the fabric of the modern neighbourhood, but constantly threatening to undermine the very foundations of all that had been built there.

Above my house on Kett's Hill lay Mousehold Heath, a forested plateau that boasted one of the most ecologically significant areas of lowland heath in southern England. Mousehold Heath was a fragment of a much larger tract of common land, covering some 6,000 acres across eight rural parishes prior to the Parliamentary Enclosures of the nineteenth century. Robert Kett's rebels had camped here. Centuries before, the last battle of the Peasant's Revolt was fought on the Heath, after Geoffrey Litster had slain the Mayor of Norwich and was proclaimed "King of the Commons", in recognition of his role in leading the resistance movement of commoners, camped on the common land of the Heath. The Mousehold Heath of today is more of a nature reserve or a park than a site of common resource extraction; the last commoners' rights were exercised here over 100 years ago (Rackham 1997, 299–301). It is managed by Norwich City Council, who at that time of my fieldwork subcontracted out some of the conservation work to The Conservation Volunteers (TCV), a national environmental charity. TCV essentially acts as a middleman between landowners and the general public, coordinating opportunities for the latter to complete conservation tasks voluntarily on behalf of the former. Volunteers gain experience, training in manual skills and ecological knowledge, and have the opportunity to socialise while doing meaningful work outdoors. Landowners obtain conservation management services at a substantially lower cost than if they enlisted a private contractor. From December until March, I spent Tuesdays and Thursdays volunteering with TCV, looking to gain some practical experience before I applied for placements in the Broads during the following summer.

On the days when I wasn't conducting interviews or exploring the wintry landscape of Halvergate or the Waveney Valley, I would join TCV Norwich's Environmental Action Group. TCV's volunteer pool was composed primarily of three demographics: retired people, students, and the unemployed.[1] Every Tuesday and Thursday at the TCV Office in central Norwich, anyone who wished to work that day would show up at 9 am, where they would help load up a minibus with tools and refreshments, before we all drove out to one of the sites TCV was contracted to manage. Mousehold was one of the more frequent destinations; there we'd cut back scrub and fell small trees. This sort of work is a vital part of heathland management; without human intervention, trees from the wooded areas of Mousehold would gradually seed into and grow up across the areas of open heath, shading out heathland species over time, until the entire area would be covered by oak woodland – a process known as succession.[2] As the Project Officer at the TCV Norwich described it:

> A lot of what we do involves cutting stuff down, simply because succession is a process of growth, and a lot of habitats that we're trying to preserve are at a certain stage in succession, and so we're trying to halt [that].

But this "cutting stuff down" aspect of heathland management on Mousehold was a source of controversy (see Figure 3.1).

On my first day on Mousehold, I was surprised to note that the biomass we removed from the heath was simply piled up and burned *in situ*, rather than harvested for fuel or as a raw material. At first, I assumed that this was done for reasons of conviviality – fires allowed volunteers to warm themselves on cold days, and were a valuable focus for socialising on site. But it became increasingly clear over time from the sheer volume of material being burned that this was first and foremost a method of disposal. When I asked why we disposed of the biomass in this way, the reason given was political rather than practical: apparently, the City Council had sought to distribute the wood harvested from Mousehold to the people of the city. Despite

Figure 3.1 Coppicing in progress.

harking back to the ancient tradition of *estovers*, which entitled commoners to harvest firewood from common land like Mousehold, this initiative proved unpopular with some members of the local community. Dogwalkers, I was told, didn't appreciate the importance of conserving heathland habitat, and disliked the removal of trees and undergrowth. A rumour had surfaced that the City Council was only removing wood from the Mousehold so it could be sold for a profit; in order to avoid further controversy, the Council wound up the scheme. I was not able to locate any official documents that could provide any further detail about this story; nor did I witness any hostility from dogwalkers personally while I was working at Mousehold. Nonetheless, the opposition was independently attested to by many of the volunteers – so it was clearly a widespread feature of their understanding of their own immediate social milieu.

The perceived hostility of local dogwalkers to heathland management, and the suspicion that had been shown towards the Council's efforts to make better use of the biomass produced by that management, was the source of considerable annoyance and frustration to those involved in conservation. But it is also symptomatic of a state of enclosure. Although dogwalkers had right of access to Mousehold, the interruption of other rights of common – to graze sheep, say, or to collect fuel – had fundamentally changed the way they saw that landscape. Unlike the conservationists – the TCV team and the council staff, who had direct, practical involvement in the management of the heath, and so knew what it needed and could reach a consensus about the best approach to managing it – the dogwalkers did not participate in Mousehold in a way that made them privy to these conditions of management. They could not see the need for felling trees periodically, and so they became hostile to the practice. Their only recourse, it was claimed, had been to spread rumours about profiteering by faceless bureaucrats, or to complain indirectly to the council hierarchy. These efforts were ultimately ineffective – in that the heathland was still being managed – and so little was achieved, from the conservationists' point of view, except good biomass was now going to waste. This outcome, it might be said, showed a complete lack of common sense.

Coppicing had revealed a difference of opinion about how the Heath should be managed, possible only through the reduction of common rights, and I was curious to discover how the conservationists felt one might go about bridging that gulf. An environmental educator, L, who at times had worked with conservation groups at Mousehold, crafting the coppiced wood into objects on site, reported that:

> Lots of people walked by, it wasn't a conversation of – "why are you cutting the trees?" – The halo comes up – "we're doing it to protect the heathland" – "yeah, but I like the trees" – And there's this sort of impasse that's reached, where they both disagree, and the dog walker walks on. Whereas, when I'm making a broom out of the birch that they're cutting down, it becomes – "What are you doing?" – "I'm making a broom." – "What are you making it from?" – "The birch" – And then, I talk about conservation, we're – "we're chopping it down, and blah blah blah, and with the stuff we've chopped down, we can make this with it, or make pegs with it, do you want to have a go?" – And they can have a go at making a peg. *And straight away, you've got a common ground, not that antagonist approach to things.*
>
> (Emphasis mine)

The practical work of crafting, for L, serves a dual purpose. At one level, it demonstrates that felling trees is not mere destruction; it is a practice that is generative, too. And, unlike the positive outcome of a thriving heathland ecosystem, a peg or a broom is a tangible, rather than abstract product; its benefits are immediate, rather than deferred. It takes butterfly transects and wildflower surveys to gauge the health of a heathland; craft objects can be held in the hand, and passers-by can "have a go" at making such things themselves. While the direct aftermath of most conservation work looks like devastation and its beneficial consequences aren't always apparent to everyone; crafting is enjoyable to watch and its products can be created and appreciated at once.

But L is also drawing attention to another, deeper process at work. The ideologies of conservation – which support the felling of trees explicitly to preserve heathland species – run into trouble when they meet those whose concerns differ. The felling of trees for conservation purposes is treated by L, here, as a partisan issue, and thus a potential source of antagonism between different constituencies. The broader resonance of this view among conservationists is evinced by the story about dogwalkers. Although they may believe their management practices to be pragmatically correct, conservationists are acutely aware that "conservation" itself is politically contested. Crafting, by contrast, is thought by L to cut across such different interests. It – "straight away" – establishes *common ground* between all those involved. A metaphor that is particularly expressive here, as it highlights both the shared understanding that is established, and the material, "earthy" character of its establishment. Crafting, then, possesses a strongly *normative* dimension. Whereas the value of conservation and the value of dogwalking rest upon distinct interests that cannot, by themselves, engender shared norms that include everyone who uses the Heath; the experience of making something establishes the opposite – *common ground* – a moral good that is compelling to everyone.

⋆ ⋆ ⋆

I've chosen to begin this chapter with three different narratives: historical accounts of resistance to enclosure; the opposition to heathland management shown by the general public, as described by land managers; and the role of crafting in cutting across the divided interests. All of them are set in a single tract of land on the edges of the Broads National Park. Each of them reveals attitudes and imaginaries surrounding the management and use of that land and its wider landscape. *The common* is one such theme, and is the one I seek to explore below. In recent social scientific scholarship, there has been an efflorescence of literature that uses the common or commons as an innovative theoretical tool, exploring political commons (Hardt and Negri 2009; Allmer 2010), urban commons (Blomley 2008; Macfarlane and Desai 2016), "paracommons" (Lankford 2016), intellectual commons (Strathern 2003; Nonini 2006) and so on. Much of this literature deals with commons of quite a different kind from the sort of commons with which Robert Kett and his rebels were concerned in my first story, that is, common *land*. The contention of this chapter is that this classic sense of the commons – as a type of relationship between a defined group of people and distinct parts of the landscape – is an important symbol in English popular imaginaries about both society and the environment, and the prototypical understanding of shared resources in general. Like the sinkholes on Kett's Hill, the trope of common ground belies many of the assumptions English people have about themselves. Opposing the ideological belief that Britain is simply a nation of individuals and families,[3] what we find is a more complex picture, in which collaboration, shared norms, and mutual understanding between individuals and beyond the intimate sphere of close kin are hoped for, forged, and collapse. As the other two stories from present-day Mousehold Heath – a common itself – demonstrates, such interactions can be fraught, and are often connected with the absence (and presence) of common sense.[4] This serves to highlight the normative quality of common sense, something that this chapter will explore in detail. As L's remarks express eloquently, the common also acts as a site of aspiration and moral creativity – a space where common ground can be plotted out, specifically through shared material experiences, materiality having been identified as a key medium through which commoning is expressed in recent discussions of the term (Jeffrey *et al.* 2012, 1249).

Pregnant with moral expectations and particular readings of history, the common carries considerable symbolic capital, even under the conditions of late capitalism – where the economic capital attached to usufructory right to land and resources is highly circumscribed, relative to the rights attached to private or public property relations (Vasudevan *et al.* 2008, 1644; Amin and Howell 2016, 2). The nature of this symbolic capital will be explored in stages. First, I will examine the role of the commons in the literature on Broadland, to establish the role of the commons as a feature of historical narratives about this region. As my first three stories indicate, these historical narratives suggest a connection between the common on the one hand and the peasantry on the other. Second, I will consider the role of the common in environmental land management in contemporary Broadland. Third, I will reflect on how the common helps us make sense of current trends in that management, in the employment structure that sustains it, and in the public understanding of rural life. The aim of this ethnographic portrait of Broadland's pasts, present, and possible future will not be to make any historical claims. Rather, the goal is to treat the discussions about the past and future of Broadland as a space where the common's symbolic role and form is expressed and articulated.

With this ethnographic course laid out, I will draw upon the extensive anthropological literature on peasant societies, particularly Pierre Bourdieu's analysis of the Kabyle (Berber) household, to argue that the concept of the common has a doxic role in English thinking about social relations among certain groups, and that it acts as a "key symbol" in English social imaginaries in general (Ortner 1973; Bourdieu 1977). This helps explain an observation made by Macfarlane (1978) regarding the relatively minimal importance of the domestic unit in English rural life going back for centuries, part of his controversial argument for the deep roots of English individualism. In English society today, I suggest, it is not the household that structures how people think about "community", but rather the *parish* – and the set of ideas and institutions derived from it – in which common land, its individual inhabitants, and their common sense ideally exist in organic unity.

Learned voices: common land in environmental histories of Broadland

The Broads' past is founded on the region's geology. Successive periods of inundation, erosion, and glaciation have layered strata of silt, marl, loess, gravel, and peat that compose the ground below Broadland; each of these types of earth, and of the processes that created them, has had its own impact upon the human exploitation and management of the region over time.

As recently as the Roman Era, much of what is now the Broads was covered by an extensive estuary system – known as Gariensis – whose principal mouth stretched from Caister-on-Sea to Gorleston-on-Sea. Gariensis was an important waterway that allowed traders access to the East Anglian interior. But over time, longshore drift created a spit of shingle across the entrance to this stretch of water as sea levels fell, and gradually the waters of Gariensis disappeared, replaced by fen and alder carr inland, and saltmarsh closer to the sea. The port of Yarmouth was built on the spit, where it still stands to this day. Over the course of the Medieval Period, large amounts of peat – over 900 million cubic metres – were extracted in the region. Local monasteries led the extraction industry, particularly St Benet's Abbey, just outside the town of Acle, overlooking the River Bure. The turbaries[5] were abandoned in the thirteenth century, due to a combination of rising sea levels that flooded the diggings,[6] and the general shortage of labour in the wake of the Black Death. These flooded basins – the broads – were soon colonised by wildlife, and their anthropogenic origins were largely forgotten until the pioneering work of Joyce Lambert in the mid twentieth century (Lambert *et al.* 1960; George 1992).

Peat digging is not the only way in which the broads have been significantly altered by human endeavour. The salt marshes have long been important as pasture, initially for sheep and horses, and then increasingly for cattle. Since the Middle Ages, tidal creeks and natural gullies have been deepened, straightened, and sometimes entirely replaced with purpose-built ditches. The result was the creation of flat, open areas of wet grassland, crisscrossed by dykes and bounded by levees. Inland, fens and reed beds were also important for grazing, as well as for the extraction of a host of other products: marsh hay and marsh litter for livestock, reed and sedge for thatching, and flag iris and bulrush for mats, lanterns, and horse collars. Invasive tree species – like alder and willow – were cleared to maintain the conditions needed to produce these important resources. Even after the large-scale medieval turbaries fell into disuse, peat extraction continued on a smaller scale, creating areas of open water that could be readily colonised by reeds and sedges. The rivers and broads were important inland fisheries and conduits for commerce, as well as a habitat for waterfowl that were shot for their meat and plumage. Distinctive local sailing vessels called *wherries* carried coal, tar, wheat, barley, clay, timber, tiles, iron, millstones, and other goods between Norwich and Great Yarmouth. To facilitate commercial travel, rivers were progressively dredged and straightened, new canals were cut, and trees were felled to allow the wind to fill the sails of the wherries without obstruction. In the eighteenth and nineteenth centuries, windpumps were installed to further promote drainage. Although these were subsequently replaced by first diesel, then electric pumping stations, the windpumps have become one of the most characteristic parts of the region's built heritage (Matless 2014, 173–182).

Although the flood plains were heavily managed for agriculture and trade, they were largely devoid of permanent settlement. The main centres of population were instead located on the surrounding upland, where the risk of flooding was lower. Channels were often cut through the flood plain to villages that were built some distance away from the main course; allowing boat traffic access via parish *staithes*[7] (Dutt 1903, 102). The upland soils surrounding these villages are some of the most fertile in the country, and while the flood plains were usually kept as reed beds or pasture, upland farms – then as now – supported arable crops like wheat, barley, root vegetables, salad greens, and herbs (Williamson 1997, 103). For 1,000 years the Broads, like the rest of Western Europe, had a largely peasant economy within which rural industries based on agriculture were fundamental (Redfield 1969, 30; Dalton *et al.* 1972, 387; Whittle 2000, 2). The landscape of East Norfolk was divided according to a manorial system, farmed by serfs and freemen, on behalf of the lord of the manor, who alone held allodial title (Stephenson 1956). Crucial to this system of production were large areas of common land, owned by the manor but to which the ordinary inhabitants of the parish had certain common rights – particularly to gather fuel, to collect fodder, and to graze livestock (Rackham 1997). Such common land was managed by manorial courts, chaired by the lord or his steward with verdicts being given by a jury of 12 local freemen. In the Broads, there were commons both on the upland and in the river valleys – in both cases, they were vital for grazing, and in the upland included more open arable than in other parts of the country (Patriquin 2004).

Such large areas of common land are now rare, with most of it having been enclosed, broken up, and sold to private landowners. Williamson traces the "long and complex history" of enclosure in Broadland, pointing out that the process was both uneven and varied. It began slowly and piecemeal in medieval times, and speeding up over the centuries, finally reaching a peak with the Parliamentary Enclosure Acts of 1797–1815, prompted by the high grain prices and an optimistic mood that encouraged capitalist farmers to invest in improvements, such as hedges, drainage pumps, and embankments (Williamson 1997, 93–98). Parishes alongside the southern rivers like the Yare were enclosed earlier and with less disruption than the northern rivers, for example, because of differences in the soils there. In the north, the upland soils were more fertile, while the lowlands were peat, rather than silt. This diminished both the incentive to enclose among the local gentry in the north – as peatier soils were less susceptible to improvement than the siltier soils to the south – and also made it harder to build a general consensus to enclose among a large community of smallholders sustained by the fertile upland soil. As such, only once Parliamentary Enclosure made the process easier, were the Northern parishes enclosed (ibid., 97). The relationship between "improvement" and enclosure is a straightforward one: the beneficiaries of Enclosure Acts would seek to make drastic modifications to the land they acquired in order to increase their profits – the most visible reminder of this in the Broads is the proliferation of drainage mills during this period – to recoup the costs of passing and implementing the Acts. By draining the land, the stocking levels of sheep and cattle could be increased, boosting productivity. Vast areas of fen, reedbed, and marsh were drained – albeit unsuccessfully in many places (Bacon 1993). But the direct impact of enclosure on the landscape was mitigated somewhat by the fact that many of the products of undrained fen, marsh, and reedbeds were still in demand. Peat was still a valuable fuel, marsh hay and litter were still important for raising animals, and reed and sedge were still important for thatching. The marshmen – those employed to manage the wetlands of the river valleys – still continued their work; what had changed was for whom they were labouring.

Over the course of the late nineteenth and early twentieth centuries, however, the Broads was subject to a drastic shift in land use. The arrival of the railway – the Norwich to Yarmouth line was built in 1844, the Norwich to London line in 1849 – coincided with the popularising of the Broads' natural amenity by a number of Victorian travel writers, such as G. Christopher Davies, E. R. Suffling, and John Payne Jeannings (Williamson 1997, 155–156). Large numbers of visitors from across the country began to come to the Broads on holiday, with the renting of pleasure boats being particularly popular. Shooting began to take its toll on bird numbers, which had fallen so low that a slew of acts of parliament were passed to protect falling populations in 1880, 1894, and 1899, while the Society for the Protection of Birds – later to become the Royal Society for the Protection of Birds – was founded in 1889 (Moss 2001, 270). Reductions in manpower during and after the First World War, as well as a sharp decline in demand for marsh hay and marsh litter due to the adoption of motor transport in cities, meant that many fens and reed beds were abandoned. Coal and gas had replaced peat; petrol and soy-feed replaced marsh hay and litter. Though the reed and sedge market continued for a time – thatch still being a common roofing material in many rural parts of East Anglia – the arrival of cheaper, imported reed from Poland put even these industries under considerable pressure. Reed-cutters today must supplement their income with other jobs, while the majority have left the trade altogether. Now left unmanaged, the fens, reed, and sedge beds were colonised by alder and willow carr, which cover large parts of Broadland today. With the introduction of the Common Agricultural Policy and the productivist mentality of post-war European agriculture, many farmers sought to deep-drain the grazing marshes on their estates and plough them up. The changes in the upland were subtler. The inefficiencies of unmechanised subsistence agriculture created "space for nature" as a matter of course – the wide margins at the edges of fields, so beloved of grey partridge, were kept as space to turn horse-drawn ploughs, for example. This meant that, prior to the twentieth century, biodiversity was a by-product of, rather than a countervailing concern to, rural productivity (Cocker 2008). With the introduction of modern, mechanised agriculture in the Broads, such inefficiencies were removed, and biodiversity began to decline (Moss 2001, 161). While the number of people employed in managing the landscape fell, the number of people living in the catchment increased substantially, with Norwich and the surrounding villages having steadily increased in size since the nineteenth century – populated by an ever-increasing number of retirees, or by tertiary-sector workers who often commute into Norwich or London for work.

This process of intensification – in terms of tourism, farming and residential development – has had a significant impact across the entire Broads catchment. The introduction of industrial fertilisers, combined with erosion caused by motorboats, and effluent from sewage treatment works catering to a booming human population, led to a significant increase in the amount of sediment, phosphorus, and nitrogen entering the Broads. This created the conditions[8] for eutrophication, with the result that the once crystal clear waters rapidly became turbid, and the once gentle, reedy shallows eroded away (Moss 2001; Ewans 1992).

These various "crises" came to a head in the 1970s (Matless 2014, 183–190). Until this period, responsibility for the management of the Broads had been distributed between a wide range of different agencies – the Great Yarmouth Port and Haven Commission (GYPHC) held jurisdiction over navigation for example, while Anglian Water monitored and enforced water quality, while the various Local Authorities were responsible for planning (ibid.). Following two studies documenting the ecological decline of Broadland – one by the Nature Conservancy Council[9] published in 1967, another by the Norfolk Naturalists Trust (now the Norfolk Wildlife Trust) published in 1976 – a consortium of representatives from different local authorities was set up in 1978 to improve the condition of the region (the Broads Authority 2014). The new non-statutory authority drew up a restoration plan, and built links with a variety of local stakeholders. One of the first major tests of the Authority was when the Halvergate Fleet and Acle Marshes Internal Drainage Board (IDB) applied to the Ministry of Agriculture for a major grant to deep drain some 2,024 hectares of grazing marsh, which would then be put under plough. The Authority objected to this and subsequent proposals to convert areas of grazing marsh to arable, citing concerns over the impact upon biodiversity. To encourage farmers and landowners to continue traditional forms of marshland management, the Broads Authority, together with the Ministry of Agriculture, Fisheries and Food and the Countryside Commission, put together a programme that paid landowners to farm the area in environmentally sensitive ways, launched in 1985 – the first such initiative anywhere in Britain. This in turn led to the Halvergate Marshes being designated the UK's first Environmentally Sensitive Area in 1987. After five years, the Countryside Commission reviewed the performance of the Authority, and concluded that a single body should be constituted with sufficient powers and resources to manage the entire Broadland landscape. This led in turn to the passing of the Norfolk and Suffolk Broads Act 1988 (UK Government 1988), which invested the Authority with the same status as Britain's National Parks (Ewans 1992).

The Broads Authority has continued to pursue its objectives since that time and celebrated its twenty-fifth anniversary during my fieldwork. Its current strategic priorities include fostering an integrated approach to the management of the entire catchment, improving efforts to conserve the region's built heritage, promoting tourism, and leading local adaptations to climate change (the Broads Authority 2013).

* * *

The contours of environmental change laid out above are agreed upon within the academic literature on Broadland. But as this canon includes naturalists (Martin George and Brian Moss), social geographers (David Matless and Martin Ewans), and landscape historians (Tom Williamson and Keith Bacon), there is a spectrum of opinion beyond the generalities sketched out above. One of the areas of most strenuous disagreement – also one of the most relevant issues when thinking about the common – is how the socio-economic changes over Broadland's history can best be characterised, particularly with regard to how these long-term changes relate to the damage to the Broads' ecology over the course of the nineteenth and twentieth centuries. Competing narratives of social and ecological change have emerged, all told through particular readings of the landscape. To put it another way, the creation of man-made lakes through peat extraction and the drainage of estuarine areas, followed by the near-collapse in Broadland's biodiversity, the eutrophication of its waterways, and the intensification of farming and development in the Broads all attest to the fact that Broadland society has changed significantly over the centuries. Where scholars disagree is in terms of how best to characterise these changes, in terms of wider trends affecting English society.

Martin Ewans' emotively titled *The Battle for the Broads* (1992) advances one particular model. He suggests that, between the medieval excavation of the Broads and the early twentieth century, there existed what he dubs "the Marshland Economy". For Ewans, the original pattern of land use described above – a combination of grazing marsh, fen, and open water, producing fish, waterfowl, peat, reeds, and other materials – was responsible for creating a patchwork of habitats and landforms that supported the Broads' unique ecology. He points to the vast catches of wildfowl and fish reported by eighteenth- and nineteenth-century writers as evidence for the thriving biota that such "traditional" land management was able to sustain (Ewans 1992). For Ewans,

> if one had to characterize in a single word the traditional way of life of the people of the Broadland marshes and villages, that word would be "self-reliance". It is true that in a primitive sense of the term they ran a market economy, in that they sold their labour, produce and artefacts, often over long distances, and in return purchased many of the necessities of everyday life. But most villages of any size had a variety of trades within their community.
>
> (Ibid., 25)

This variety of local expertise was nourished by a landscape that provided a wealth of different raw materials close at hand. Even if commerce was a feature of traditional Broadland life, Ewans would argue that its influence was initially only slight. And so, "it was only with the improvement of communications with the outside world and the immense increase in mobility of people and goods that the traditional Broads Economy began to break down" (ibid., 25). This depiction of Broadland society – as largely static, in harmony with nature, and remote from the concerns and pressures of wider English society, yet doomed to ultimately be overcome by them – clearly owes much to the romantic descriptions of rural life authored by earlier pseudo-ethnographers of Edwardian Broadland, such as Arthur Patterson (Tooley 1985), William Dutt (Dutt 1903), or Richard Lubbock (Lubbock 1845), and photographers such as Peter Henry Emerson (Emerson 1885) or William Henry "Eugenia" Finch (Austin *et al.* 2011). Reflecting prevailing attitudes in British folklore studies at the time (Bronner 1984), and in Boasian anthropology (Albright *et al.* 2002, 46), as well as invoking the recurring ethnographic trope of the "isolated rural community" (Rapport 1993, 32–33), the prevailing aim of such work was to document Broadland's "traditional ways" – understood as being both largely static, and highly parochial – before they disappeared in the face of inexorable modernisation, of which enclosure of common land is a critical feature. Emerson, for example, concluded grimly that "when the land shall all be built upon and enclosed, and the peasant is no more, then may old England go grovel before the world" (Emerson 1893, 27).

An alternative prospect is advanced by Tom Williamson, in *The Norfolk Broads: A Landscape History* (1997). *Contra* Ewans, Williamson is critical of attempts to invoke "tradition" as an analytical tool, arguing that environmental change has been pretty much a constant in the Broads, reflecting complex interactions between societal, economic, and environmental factors that are all in a permanent state of flux. For Williamson, tradition is a trope that plays an important, but often quite negative role in both public policy and popular discourse:

> [T]he Broadland which we enjoy today is not some timeless entity but an amalgam of numerous phases of development, many of no great antiquity. Once the complexity of history is thus denied, "tradition" becomes a guiding principle in landscape management but, unquestioned and undefined, this term is rapidly emptied of whatever meaning it once possessed, and the real world of life and labour is forced to masquerade as historical pastiche.
>
> (Williamson 1997, 164)

He argues that even "aberrant" features associated with the "decline" of Broadland – such as the increasing size of Broadland villages, riverside chalets, and other developments supporting tourism – are as much part of the landscape as anything else, and that many of the features of Broadland that are treasured today – such as wind pumps, easily navigable rivers, or even the broads themselves – are the products of industrial exploitation that would, when they were first created, have horrified people today. The creation of common land as part of the manorial system, and its eventual enclosure as peasant agriculture faded away, is for Williamson but one thread in the diverse tapestry of Broadland's landscape history.

Such criticisms of the traditionalist position are echoed by Brian Moss (2001), in the postscript of his limnology of Broadland. *The Broads: The People's Wetland* (2001) is dedicated to a causal analysis of ecological decline in Broadland – with a particular focus upon eutrophication. In the postscript, he presents two alternative visions of the future of this landscape – one dystopian, one utopian. In the former, Moss envisions a sequence of events where the Broads Authority is replaced by a private company that turns the broads themselves into a leisure space reserved for the very wealthy. Sea-level rise is kept back through embankments, wild animal and plant populations are artificially preserved through genetic manipulation and intensive husbandry, and local communities are replaced by Authority employees dressed in period costume (Moss 2001, 350). The wider British landscape, meanwhile, is consumed by ever-greater intensification – of industry, agriculture, and urban development. Moss's utopian alternative assumes a global shift towards an economic system where costs reflect environmental impact, rather than exchange value. This provides a much-needed boost to primary and secondary industry in Broadland (cast as a more sustainable alternative to global resource flows), which in turn prompts a return to forms of land management that support greater biodiversity (Moss 2001, 356). The Halvergate Triangle and the Upper Thurne Valley are left to the sea, providing habitats for migrating birds and space for boating and watersports, while the rest of the broads are managed to meet both human and non-human needs through local industry. Adopting an imagined retrospective, Moss goes on to say that:

> because Broadland had now become a working landscape again, there was much to be seen and a real sense of continuity with the past, which

> had, paradoxically, been interrupted in the doldrum years between 1960 and 2025. Then its management had largely been one of frantic attempts to preserve items of interest against forces that were inexorable and destructive.
>
> (Ibid., 360)

Although Moss here confines these disruptive forces to the period from the late twentieth to the early (and imagined) twenty-first century, other historical accounts and events draw attention to conflicts over resources reaching over much longer timescales. They also highlight the relationship between those who work in the Broads, and commons – something integral to Robert Kett and Geoffrey Litster's rebellions. These rebellions were resistance to a trend of increasing elite control of the landscape, a trend that intensified over the course of the eighteenth and nineteenth centuries, referred to as the *Enclosure Movement.* Francis Pryor defines enclosure as:

> a way of partitioning the landscape to indicate that particular fields or farms are owned by certain individuals or estates, who generally possess written title to them, in the form of deeds. Enclosed land, unlike Open Fields (*sic*), commons, heaths and moors, cannot be owned by several people.
>
> (Pryor 2011, 380)

Enclosures organised by local landowners – known as enclosure by agreement – took place across wide swathes of the British Isles before the Enclosure Movement, although much of the Midlands, the South, and East Anglia – such as the North Rivers – remained unenclosed until the passage of some 4,000 parliamentary acts passed between 1750 and 1830 (ibid., 465). These acts sped up enclosure considerably, and formalised the process of enclosure by agreement, appointing commissioners to survey each parish, ensuring that public rights of way were maintained and that smaller landowners were not penalised. In practice, however, the Enclosure Movement concentrated landholdings into fewer, larger estates, especially as the petition to enclose only needed the support of those who owned the majority of the land in a parish, not the majority of the population, or even a majority of the landowners (ibid., 465–468). The Enclosure Movement ensured that many manorial commons were converted into private freeholds, fenced, intensively stocked, and often ploughed up by their new owners (Williamson 2000, 1997). Numerous scholars have framed this in terms of destructive class or colonial warfare; Polanyi called the enclosure of the medieval open field system "a revolution of the rich against the poor" with the consequence that "the fabric of society was being disrupted" (Polanyi 2001, 35; Durie 2011; Wright 2016, 90). That fabric is often shown in a positive, communitarian light, which, regardless of its accuracy, is revealing of the attitude of the historians who describe it. Patriquin, for example, states that:

> this [common, open field] system meant that farmers had to cooperate with each other extensively. And in a communal environment, there was little room for individuals to force changes in agricultural techniques on the village. Alterations could only be made if they met with the approval of most members of the community.... The community, through the manor court, established substantial bylaws with penalties to maintain efficiency and fairness.
>
> (Patriquin 2004, 202)

The precise contours of how this communal parish system was dismantled have been mapped by a number of Marxist historians, such as Robert Allen (1992), and E. P. Thompson (1991). This view is perhaps best characterised by Thompson, who in the same breath punctures any romantic idea that commons management was some sort of primitive communism, saying that: "Common right, which was in lax terms coterminous with settlement, was local right, and hence was also a power to exclude strangers. *Enclosure, in taking the commons away from the poor, made them strangers in their own land.*" (Thompson 1991, 184, emphasis mine), referring to enclosure elsewhere as "plain enough case of class robbery" (Thompson 1980, 237; Frake 1996, 101; also Vasudevan *et al.* 2008, 1641–1642). Ewans describes the aftermath of this estrangement in terms of mass emigration from rural communities (Ewans 1992, 12). The injustice and anger caused by enclosure was not just in retrospect; it helped fuel the rise of Chartism, workers' movements, and other left-wing political projects, in the increasingly crowded and dirty cities (Miller and Ayer 1981, 379; Rosenman 2015), a response to what Karl Marx famously described as "primitive accumulation" (Marx 2000, 521–523). Moss also stresses the harm that the enclosure and agricultural improvement associated with primitive accumulation did to the situation of the rural poor, echoing the grievances of Robert Kett:

> The enclosures, though they had increased production by encouraging landowners to invest in new methods and machinery and to drain wet land, at first brought work to the glut market of labourers at the end of the wars, but then reduced them to servitude. Common rights – gleaning the fields, common pasture on the fens, free fuel and fish – disappeared. Poaching became a way of survival and was severely put down by Game Laws.
>
> (Moss 2001, 162)

Drawing on the work of Glassman (2006), Irvine *et al.* contend that enclosure can productively be viewed as a continuing process in the English landscape, where:

> children find little place for themselves within the designation of spaces for development and of spaces for nature; in as much as they experience

> "naturedeficit" it is precisely because they find themselves in the midst of a land-use conflict. Such conflicts reveal the ongoing character of enclosure in already enclosed spaces, particularly in the wake of the encouragement of housing development and a new commodification of the countryside for leisure.... Yet, the rich examples of reappropriation we have found in our East Anglian research suggest that inherence and inclusion in local environments matter to children.
>
> (Irvine *et al.* 2016, 949)

Much as East Anglia's children still discover opportunities to engage with their surrounds, so do the adults who write about it. In many respects, the utopian vision that Moss constructs constitutes a manifesto for the reappropriation of Broadland by restoring the character of the Broads as a "working landscape" to which the entire local community had clear rights; an ideological move that would allow it to live up to Moss's titular epithet for it as "The People's Wetland".

But the singular narrative of enclosure and alienation isn't the only way in which social change in Broadland has been represented. Williamson's attentiveness to the complexity of the area's landscape history is taken into the social domain by David Matless (2014), whose cultural geography stresses both the multivocality of the region and the intimate connection between culture and landscape – within which an interweaving of different narratives in art, poetry, photography, and literature catches animals, plants, people, and technologies in the wetland warp and weft:

> Parallel regional matters of hydrology and identity shape Broadland, whether in the maintenance of grazing marsh as an iconic regional landscape, contests over rights of navigation, or the defence of fen and reedbed as home for regional fauna and flora, against both human reclamation and natural succession.
>
> (Ibid., 4)

Matless tracks the unfolding of discussions around personal conduct, animals, plants, causes of ecological decline, and much-loved local icons – wherries and wind pumps (ibid., 5–6). Of particular relevance to our discussion here is Matless's third chapter, in which he undertakes an analysis of local norms tracing the often vituperative disagreements between different groups in the region (Matless 1998, 28, 2014, 55–96). One such normative trajectory emerged during the early twentieth century, when:

> Broadland discovery [as a literary genre] was preoccupied with authentic regional conduct, on the part of inhabitant as well as visitor, working long-established conventions of seeing the rural worker (for Emerson the "peasant") as both the bedrock of England and something internally exotic in a modernising county, at once English essence and internal

> other to urban modernity.... Working lives, customs and dialect signalled cultural value, though their pursuit could be for varied political ends and different cultural effect.
>
> (Matless 2014, 64–65)

From this perspective, as Matless goes on to explore, the damage done to the Broadland landscape in the later twentieth century was seen as a consequence of poor conduct by visitors: "Eutrophication ... suggested Broadland destroyed by excess, matching for some an equivalently excessive post-war human conduct, abundance all round destroying the region" (ibid., 184). As Matless describes, this is only one of many collections of "voices" that speak about Broadland, but it is the one that is most relevant to our present discussion – because it claims that the valorous, ecologically sound work of Emerson's peasants, proper to the Broads as a region with a common set of norms, was displaced by the damaging presence of tourists, who had no real connection to the landscape, or any common sense of how they should behave there.

Working voices: "bad farming", tidiness, and the balance of contemporary rural life in Norfolk

Among the reed beds one sweltering August day, I was helping rake up cut reed with my line-manager, O. O had worked in rural industries his entire career, having begun as a stockman before transferring into conservation. In addition to being very knowledgeable, he was happy to talk about what he knew, so our time together was an excellent opportunity to collect material for my thesis. On that occasion, we were talking about what motivates people to volunteer on nature reserves. O believed that people derive a sense of meaning from work, and that without that they tended to become listless. Work was, therefore, vital to people's self-esteem. O explained further, as he leant on his muck fork, that it was for this reason that so many retired people seek out voluntary work after they retire: "They want something to do." O referred to one of the volunteers he'd worked with as example – having retired from the police force, with a good pension, and no need to work, this man had nevertheless got a job with Norfolk Wildlife Trust, did a degree in Conservation Science, then got a job working with the RSPB. But O also felt that this drive to seek meaning did not prove to be sufficient encouragement for people to work if they didn't need to do so at all. People were, in O's view, fundamentally selfish and lazy – these being part of "human nature". It was only the threat of destitution, enforced by the rule of law, that kept economic life going. Within a short time, O had voiced two prevailing, somewhat contradictory, assumptions about human nature that I found to be highly prevalent in my fieldsite: on the one hand, that human beings feel an inherent desire to be "productive", and that humans need to be encouraged to work, otherwise they would remain idle.[10]

Similar themes emerged in a discussion I had with the aristocratic owner, R, of a large estate in Broadland. For him, a culture of reciprocal obligation was what rural life was all about, something that he felt policymakers and urbanites couldn't fully appreciate. Driving past hedges and copses of trees that sheltered feeding stations for pheasants, we toured the fertile upland overlooking the floodplain of the river below. He described the maintaining of this landscape in terms of "public duty" – trimming hedges was something that he wasn't paid to do, but "just did". But there were limits to how far this public-spiritedness could extend, and where self-interest would intervene: "At the end of the day, I am running a business." As we drove by a field, we saw two grey partridges in a fine example of their habitat: broad banks of long grass, with standing perennial weeds in the field itself. My host referred to these wide field margins euphemistically as "bad farming": farmers seeking to maximise productivity today would normally plough them up to make space for more crops, and wouldn't allow weeds to grow in the open field over winter. Providing financial incentives for such measures is a key element of European Union-funded agri-environment schemes, a fact that merely serves to underscore the norm of maximising the area under the plough (Hackett and Lawrence 2014, 8). An estate-owner, C, with whom I spoke on a later date agreed – saying that many farmers had a strong desire to "tidy up" their fields, to meet certain common standards of what a well-managed landscape should look like (See Figure 3.2).

Figure 3.2 A depiction of weeds being allowed to grow amid the stubble on a harvested wheat field. Though great for wildlife, such practices are widely rejected by many farmers in Britain.

These vignettes capture many important, yet tangled features of rural industry – particularly farming – in England today. First, the desire to increase profitability by ploughing larger areas and boosting crop yields – sacrificing one sort of margin for another – reflects the prominence of capitalist logic within contemporary British agriculture. The farmers I spoke to, like R, would emphasise that although they cared for the environment, often very deeply, they were ultimately "running a business" – which came with certain financial considerations. Self-interest, as O pointed out, was seen as fundamental. Burton emphasises the connection between the productivist ethos of "good farming" and the desire to maximise the nation's food supply (Burton 2004, 195–196), but the farmers I spoke to stressed their status as individual "businessmen" in equally strident terms – the two concerns are, I would suggest, mutually reinforcing. There was a spectrum of opinion, with some landowners stating emphatically that stewardship of the rural landscape was more important than profit; for others the bottom line was the bottom line. But for all those involved in managing the land, getting the margins right mattered.

The second point is that intensive land management isn't solely a matter of economics; the concern for "tidiness" reveals a distinctly aesthetic – even moral – dimension. As Burton observes, "For many farmers [productivism] represents a picture of good farming practice, displayed in a manner that enables the farmer to obtain social status and recognition amongst their peers[11] as a 'good farmer' and to judge the credentials of others" (ibid., 208). Hall points out the connection of this attitude to the broader policy environment, saying that

> for over thirty years, the CAP has generated (and subsequently reinforced) deeply internalised norms of "good farming" that underpin how farmers see the world and feel about their contribution to society … this worldview is based on internalised moral responsibilities of producing necessary food for hungry people.
>
> (Hall 2008, 25, emphasis mine)

Although this disposition was ingrained in the farming community by the postwar domestic policy agenda and the Common Agricultural Policy (Dobbs and Pretty 2001, 2; Burton *et al.* 2008), it has old roots. Richard Irvine and Mina Gorji have described how a similar aesthetic was also a crucial discursive trope in the drainage of the Fens – the Protestant valourisation of labour motivating a desire among landowners to make the "idle" wetland into productive fields, remaining compelling for contemporary farmers who are critical of re-wetting (Irvine and Gorji 2013).

These two points are important, because they indicate the ways in which land managers – whether they are conservationists like O, owners of estates like R, or farmers like C or those studied by Burton – situate themselves in different positions along the axis of self-interested individualism and shared, *common* norms. Although farmers regarded themselves as individuals, the fact that "good farming" is tied to collectively understood physical signs left by working the landscape in certain normative ways underscores the point made by L above – that working together is deemed to be generative of normative social relations, and thus ends up having this effect. Farmers, therefore, do not readily conform to an ideal type of capitalist entrepreneur, but are informed by a range of dispositions, inculcated through their experience of being observed, and observing the effect of other farmers' labours on the landscape. Even if they could earn more from an agri-environment scheme, for example by behaving differently, the power of collective work is such that farmers' habitus mandates "tidiness" instead.

Rural sociologists emphasise the importance of habitus to farming, through which work "etched" on the landscape through practice in turn reinforces that habitus (Hall 2008, 36; Burton 2012, 54). I will return to this below, but here it is sufficient to juxtapose these observations with L's comments above – that working together sustains common ground. Collective assumptions about what "good farming" involves, generated through working together, clearly does not displace self-interest or individualism. The overriding impression I had was that most land managers saw their working lives as being a balance between these two competing, yet equally natural interests. A dairy farmer I spoke to, S, made this explicit when I visited her business. She ran a small dairy farm with her father and other members of her family. Faced by falling milk prices, they had struggled to make a living with their small herd, and so S had decided to raise the price they could expect for their product by using their milk to create artisan cheese, which she now successfully distributed to farmers' markets and department stores across the country. Despite possessing strongly entrepreneurial qualities, S repeatedly emphasised to me the importance of maintaining a "balance" – between the "traditions" her father embodied, and business concerns. The essence of the balance, as S saw it, was the question "Are we having a nice time?" – "we" meaning the humans, the animals, and the land. This balance was inimical to profiteering (which would cause the land or the animals to suffer) but also did not sit easily with the regulatory burden established by public bodies. Of the Common Agricultural Policy – of which more will be said in Chapter 5 – she remarked: "How common it is, I just don't know."

Concerned voices: current trends in Britain's rural economy

The comments I've discussed above were made at a time when rural industry in England was undergoing dramatic changes that have direct implications for the ideal of a balanced relationship between self-interest on the one hand, and the common on the other. At the time of writing, a tiny proportion of English people – even in rural communities – work directly in rural industries: just 1.3 per cent of the UK population in 2012 (UK Government

2012, 10), part of an ongoing pattern of "labour shedding" from the rural economy (Lobley *et al.* 2005, iii) and consolidation of rural businesses. I was shocked by how few farmers I met, even at the Norfolk Farming Conference and the Anglia Farmers Tent at the Norfolk Show. A great many "rural" people whom I met were not farmers at all, but waged professionals – contractors, land agents, or estate managers – employed to farm land with which they had no long-standing relationship, on a purely contractual basis. Many were members of families who had once farmed small estates they'd owned, but had subsequently sold, usually retaining a small fleet of agricultural equipment that they would use for contracting; or they had retained ownership of the land, but rented it to another, more commercial farming operation. Since the 1960s, more and more farms, particularly small family-owned operations, have been aggregated into fewer, larger estates (see Strang 2004, 14). Since 2000, for example, the proportion of English farms below ten hectares fell from 33.8 to 12.5 per cent, while the proportion of farms above 100 hectares increased from 18.3 to 39 per cent (Winter and Lobley 2016, 20). In their 2016 study of a single West Country parish, Winter and Lobley found that of 26 farms which were operating in the district in 1941, 16 had ceased to exist – having sold their land to other operators. Only one of the remaining ten farms remains a family operation, with the others either letting out their land to other farmers, being run as "hobby" farms by wealthy owners, or becoming dependent upon non-farming earnings (ibid., 25). Although small family farms remained a robust component of land management and rural habitation into the twenty-first century (Lobley and Potter 2004, 508), it is also equally clear that rentiership is becoming progressively more important, posing distinct challenges for smaller operators (Winter and Lobley 2016, 34–35). As a result, many small landowning families – whose ancestors might have initially benefited from enclosure – have over time followed the rural poor of the eighteenth and nineteenth centuries, off the land and into the wage economy, or even out of agriculture altogether.

Such concentrations of landownership aren't inimical with "tradition" in Broadland. In the Halvergate Marshes, for example, most of the land is rented by graziers, who bid for summer lets at an annual auction in March at the Bell Inn in Saint Olaves, beside the River Waveney – a practice that, in its exercise, had become something of a tradition. According to a local historian I spoke to, Norfolk had "always been a county of big estates", and so patterns of renting and waged employment in rural industry are by no means a novelty. But even the traditional landed estates are not unaffected by the present economic trends. One estate employee explained that:

> the generation running these estates now mostly have business interests in London and elsewhere. The estate is, that landed tradition, they go there for the shooting season, they might be there once a month, it's not necessarily their main residence – doesn't necessarily make them absentee landlords. They will have staff in place to run their estates … the

> overarching strategy is [the owner's], but the guy at the top does not do it himself.... There are very few that have inclination to go to college and do the training, to give them the capabilities to run their estates. So they pay someone else to do it.

Large estate owners of the present generation, like R, retain a deep sense of personal commitment to the management of their land, and occupy a strategic role within their businesses – unlike true "absentee" landlords, who demonstrate neither. However, it is important to note that they are already heavily reliant on other business interests, often pursued by other members of their family. As my informant went on to explain, the close relationship between estate owners and their estate was probably going to change for future generations as a result: "I think, if not the generation immediately coming up, then certainly the one below them ... [sighs] the cynic in me thinks things will change.... You won't have the same personal interest, that personal investment in these estates." The children and grandchildren of the current generation of gentry and aristocrats simply have different aspirations to their fathers and grandfathers.

The effects of these shifting expectations – even after a family has sold its holdings or moved off the land – emerged during a conversation I had with Brian Grint, a local historian from the Broadland town of Acle, who was born and brought up in the nearby village of Halvergate. Brian possesses a vast amount of knowledge about the local area, including the marshes of the Halvergate Triangle (see Grint 1984, 1989), and offered to give me a tour of Acle and answer any questions I might have about the history of the surrounding landscape. Brian was a leading figure within Acle's Historical Society, and explained the past of various local landmarks, old and new – the church, the crossroads at the centre of town, and the site of an old market where cattle were once bought for grazing out on the marshes, but where a Budgens now stood. The Historical Society had added various new features, too – including a wooden causeway built by local families, and other footpaths. Brian was justly proud of what he had achieved – when I asked him if he felt the Society had been successful in encouraging interest in Acle's past, his response was unequivocal: "Absolutely."

By this point in my fieldwork, I had read about the disappearance of the marshmen – the cadre of rural workers who tended the wetlands of the Broads in the nineteenth century. I asked Brian where the marshmen's descendants were now, and he pointed proudly at himself and said: "They're here!" Although the marshmen's way of life had declined, their descendants still remained in the area, he went on, albeit in different professions. Later, Brian said that his daughter – a photographer and producer – was based in Norwich. Although she loved the countryside, he told me she planned on moving to North Norfolk, if and when she left the city.

Brian's family history, I suggest, is illustrative of the changes affecting rural people across Broadland. In our collective imaginaries at least, Brian's ancestors were deeply rooted in the local landscape, having even leant their surnames to the marshes they once tended.[12] Brian himself, however, took work as a film producer, and moved from his natal village of Halvergate to the nearest town of Acle. His connection with the local area was expressed not through the tending of the land, but through his personal interest in local history, pursued in his spare time – the majority of his working life being spent elsewhere. His daughter relocated to Norwich, travels the world for work, and has no plans to return to Broadland. Though the connection Brian had to "his marshes" was no less strongly felt for all this, the economic transformation in a matter of generations is nonetheless profound, and not everyone was as unperturbed as he was by these changes. As Anderson points out, "Temporariness forms a barrier that significantly impacts on the type of interactions that occur between working holidaymakers and long-term residents" (Anderson 2016, 6) – a dynamic that also applies, I suggest, when comparing recent arrivals to those whose families have lived in a place for generations, and interactions with the landscape as well as with one another (see also Strang 1997, 288). As with the owners of landed estates, those personal connections are felt to be disappearing among younger generations, whose aspirations are tied to a far less local, and increasingly individuated, set of priorities. The balance between the common and the self-interested individual is continuing to shift.

Even for those who remained working the land, this shifting of the balance had profound effects. My contributors stressed the extent to which farms themselves have been forced to diversify, becoming commercial businesses first and foremost, with farming being only part of an extensive portfolio. J and D – two brothers with a family history in the agricultural sector – began as solo farmers, before setting up a partnership in the North Rivers. This partnership both farmed its own estate, while also providing contracting services for other landowners. This change, J and D said, was typical of the direction of agricultural sector as a whole, as it allowed farmers to benefit from economies of scale, and boosted efficiency. The meteoric rise of Anglia Farmers, an agricultural bulk-purchasing group based near Norwich, also reflects this trend. Formed in 2003 from a merger of two, older buying groups, Anglia Farmers now purchases 10 per cent of the UK's total farming inputs, and has an annual turnover of £250 million. Its members, the Chief Executive Clarke Willis proudly told me, farm 60 per cent of Norfolk. In addition to the very existence of this purchasing group being a case of consolidation, Clarke also mentioned other important trends in the sector that indicated the commercialisation of farming. Supply chains to consumers were becoming shorter and more integrated, with supermarkets showing a greater concern for maintaining their sources of key products than previously. Clarke also felt that most farmers, at least in Norfolk, would welcome the disappearance of the Common Agricultural Policy (CAP), as it "distorts the market".

These trends – towards consolidation of land ownership into progressively larger holdings; a reduction in the size of the rural labour force; commercial consolidation due to market forces; and a gradual lessening of interest in land management by land owners – was observed by almost everyone I spoke to. Certain authors and commentators – such as Kevin Cahill, or Carol Wilcox – have stressed the inequity of land ownership in the UK as an overlooked cause of contemporary social problems that allows for the transfer of vast sums of money in the form of government subsidy to the country's wealthiest citizens (Cahill 2002; Adams 2011). I would suggest that the inequity of land ownership is part of a much larger, cultural change. Fewer and fewer people are participating directly in working the land; and what's more, the *nature* of that work is changing. Rather than being overseen by family farmers or by hereditary aristocrats, and performed by the majority of the local population, an increasing amount of rural work is being done by a small number of professionals, or volunteers motivated by personal interest. Rather than being a way of life, then, land management is becoming either a job, or a hobby, for an individually interested few. Under such conditions, people draw upon their basic assumptions about the commons, community, and the countryside in order to make sense of how their lives and surroundings are changing, even if their daily practices are structured according to self-interested, individual, and often commercial interests. The gradual collapse of the proportion of the population involved directly in rural industries, and the professionalisation of this sector, has meant that the forms of knowledge attached to them have, in turn, started to retreat from popular consciousness. As one land agent, whose family were farmers, explained to me, this meant that fewer and fewer children gained an interest in agriculture as a career; in her words, "We're losing our young people." These sentiments were echoed by D, who claimed that education had "missed out a whole generation of children" with respect to farming. As another farmer I spoke to put it, "The people who want to work on the land are either farmers' sons, or they're special needs." English agriculture today has become increasingly specialised and professionalised – a technical practice as remote from the working lives of most British people as particle physics, and just as poorly understood.

This was something Clarke Willis wanted to stress, too. Half-way through our interview Clarke asked me, with an air that would have been pointed if he wasn't so personable: "How long is the supply chain for milk?" I glanced at the jug of milk on the table beside our cups of tea. I searched the air for numbers. "A week? Maybe less?" Clarke's eyes sparkled in triumph. "Four years! It takes about that long to rear a milker; it would take this long to restore our milk supply if all our cows disappeared!" Clarke had vigorously demonstrated the ignorance of urbanised, metropolitan academics like me about the basic matter of how one gets from field to fork, starting from scratch. Under Clarke's leadership, Anglia Farmers has attempted to address this knowledge gap, organising public outreach events like Open Farm Sundays. But the very fact that such events are deemed necessary indicates that a fundamental shift has taken place. What once was common knowledge – knowledge found through the work that everyone would do in daily life – has become much more specialist. This trend was perhaps highlighted best by my experiences engaging with people at the opposite end of the scale to the sort of businesses interviewed so far.

B and K, two members of a farming cooperative at which I worked, mused about how few British people felt attracted to the notion of "peasantry". K contrasted this against the French attendees at Farm Hack – an event they had attended recently – who actively embraced their identity as a modern, rural peasantry, devoted to self-sufficiency, community, and sustainability. This lack of popular investment in agriculture was echoed by a farmer I spoke to, who had attempted to offer his expertise to a community-supported agriculture (CSA) project locally. But the experience had shocked him:

> "I had to teach grown men to use a spade and dig a hole, and teach them how to use a hoe. I thought there would be queue of people wanting to help build things, make things, do things, and be involved with the infrastructure of a CSA, and there weren't ...".
>
> "No?"
>
> "Nope. They weren't there. They're not there."

The lack of interest in community supported agriculture, and the low level of skill among those who were interested, was claimed to be indicative of how rural, "traditional" culture was in a state of collapse. We see, in these words and anxieties, echoes of Emerson's lament over the fate of the Broadland "peasant".

These trends in the rural economy had broader social consequences that were frequently lamented by local people. Broadland's position between the three large urban centres of Norwich, Great Yarmouth, and Lowestoft has meant that many villages here serve as "dormitories" for affluent commuters or second-home buyers, while poorer families from urban areas are relocated to villages with social housing.[13] One of my contributors, a former reedcutter, had to move away from his idyllic natal village of South Walsham in the Bure Valley due to rising house prices. He explained that Ludham – a more distant village to which his entire family had relocated – was better because it "still had a bit of a community". Another contributor, a resident of the village of Reedham, pointed out that the majority of locals commuted to Norwich for work. She explained that:

> The village is quite split now.... We have got what I call "proper Reedham people" still, you know, generations that have been to this school, that we've still got here, but we've also got other, new families, because when they built new houses in the village here, they're what they call social housing....

W, who lived in Southwold, complained bitterly that the pretty seaside town was almost completely dominated by Londoners buying second homes.

Young families who had worked on local farms, she said, had been evicted from a row of little cottages on the outskirts, which had then been sold for vast sums of money to incomers who were hardly ever there. The situation had grown so bad, W said, that at Christmas, she recalled whole streets where only a third of the houses were occupied. Local residents were keen to link this process by which local people were displaced from local housing to a "lack of community". The ethos of mutual aid and communal sentiment invoked by R above was, in the eyes of many of those who lived in Broadland, lamentably absent. Just as rural industries are becoming increasingly integrated into the capitalist economy of Britain and being drastically consolidated as a result, so are the villages where its peasant workforce once lived becoming increasingly interconnected with the wider, urban-dominated housing market. Individuals and families from outside the area, be they wealthy commuters or the recipients of social housing, are depicted as threats to the ideal of village community rooted in the local landscape that it manages.

Analysis: work, common land, and the process of enclosure in Broadland

The ethnographic survey I have provided above deals with three main types of material. The first – the Broadland literature composed by intellectuals – is concerned first and foremost with constructing narrative pasts about this landscape: how it was formed, managed, and came to be the way it is today. Following Matless, I would suggest that what emerges from these accounts are specific norms, with the work of rural labourers of the past being deployed as a standard against which other groups and forces – including modern selfish individualism – are judged. The second source – conversations I had with present-day land managers about their work – shows that work has a normative force for them too, inscribed in a tidy landscape with all the right margins. But we also find a more nuanced set of attitudes. The individual, here, is not inimical to this working ethos, but is something that must be held in balance with it. Both common norms and individual interest are seen as natural parts of rural life. But when we turn to the third set of materials, broader economic trends, and people's expectations about what these trends mean for the future, we see that this balance between self-interest and common-ground is felt to be shifting in favour of the former.

There are various points of contact between these three sets of materials – particularly the scholarly literature with which I began – and anthropological perspectives on cultural life and the environment. "It's complicated" has been referred to as the anthropologist's favourite answer to any given question (Kaplonski 2015) – and this same sensitivity to the specific is a clear theme in Frake's, Williamson's, and Matless's writings. Matless's approach – treating the cultural landscape of Broadland as a colloquy whose diverse meanings can be traced by the attentive participant, is profoundly ethnographic, and clearly owes a considerable debt to the interpretive anthropology of Clifford Geertz and Victor Turner, albeit with a view to, in the words of Patricia Price, "track[ing] culture, instead of trying to cage it" (Price 2010). As a portrait of the cultural life of a watershed, Matless's work has close parallels to Veronica Strang's survey of the meaning of water in the Stour Valley (2004). Here, it is possible to build upon Matless's work by examining the relationship of these cultural processes to economic and managerial processes that he neglects, but that Strang explores (2004, 129–192). In criticising the conservation movement's protectionist agenda, Moss echoes similar concerns voiced by anthropologists, who have criticised the way in which national parks in colonial contexts have often been protected through the wholesale eviction of local populations (Robinson and Redford 1991; West and Brechin 1991). As with Ewans' writing, and the salvage ethnography of Lubbock and Emerson, the more recent phase of environmental history of Broadland therefore intersects closely with wider concerns within anthropology. But here I'd like to focus upon a comparison of my material with the anthropology of peasant societies.

The anthropology of peasantries is an appropriate source of insight for two reasons. Peasants feature within the discussion about Broadland's past, and so comparing these discussions to ethnographic descriptions of peasant societies is a logical step. Second, both the qualities attributed to Broadland's peasants – that their work invokes a strong normative framework, and that through this work they are connected intimately with their surroundings – are key traits identified in peasant societies by anthropologists, and, as we have seen, these are also key features of common sense. This perspective on the peasantry is perhaps exemplified by Redfield, who states:

> One sees a peasant as a man who is in effective control of a piece of land to which he has long been attached by ties of tradition and sentiment. The land and he are parts of one thing, one old-established body of relationships. This way of thinking does not require of the peasant that he own the land or that he have any particular form of tenure or any particular form of institutional relationship to the gentry or the townsman … [only that] they have such control of the land as allows them to carry on a common and traditional way of life into which their agriculture intimately enters, but not as a business investment for profit.
>
> (Redfield 1969, 27–28)

Redfield stresses, here, that ownership is not what defines the peasant, but rather "control of the land", i.e. the fact that they manage it. This close relationship gives the land an almost sacred significance in peasant societies (Macfarlane 1978, 23–24). But this distinctive, hallowed working relationship between peasants and the land is disrupted by the logic of the market:

> It is the market, in one form or another, that pulls out from the compact social relations of self-contained primitive communities some parts of

> men's doings and puts people into fields of economic activity that are increasingly independent of the rest of what goes on in the local life. The local traditional and moral world and the wider and more impersonal world of the market are in principle distinct, opposed to each other, as Weber and others have emphasized. In peasant society the two are maintained in some balance; the market is held at arm's length, so to speak.
>
> (Redfield 1969, 45–46)

The penetration of wage labour – so ubiquitous in Broadland's rural industry today – "destroys the economic logic" of the peasant class, while industrialisation removes its *raison d'être* entirely (Franklin 1971, 99–102). Wolf and Lewis view the trajectory of peasant societies exposed to economic development similarly, although emphasising the role of private ownership, rather than wage labour, and the technology of the plough:

> [In Mexico] once private ownership in land allied to plow culture is established in at least part of the community, the community tends to differentiate into a series of social groups, with different technologies, patterns of work, interests, and thus with different supracommunity relationships.
>
> (Lewis 1951, 129–157, cited in Wolf 1956, 1070–1071)

Dalton develops this point further, concluding that:

> perhaps one may sum up the economic changes that transform the peasantry during the late stage of development and modernization by saying that rural households and villages become more dependent upon market forces and governmental services external to the village. The Gemeinschaft qualities of local life carried over from the feudal period and early modernization are further weakened as villagers and villages become less and less isolated from urban and national life, and more and more dependent on their transactions with and participation in institutions outside the village.
>
> (Dalton *et al.* 1972, 397)

Dalton provides a portrait of socio-economic change that resonates strongly with the experiences of those living in my fieldsite. This suggests that, from the perspective of the environmental historians of Broadland, the status of contemporary farmers and rural workers could be deemed that of "postpeasants" – people whose way of life has undergone a shift from that of a "true" peasantry, under the influence of modernisation and capitalisation (Gamst 1974, 11).

Although they strive for a "balance" with market forces, as we have seen, the connection between common and market sensibilities is anything but "arms-length" in Broadland. Individual self-interest has become part of local "common sense", as David Harvey would contend (2007, 3, 116–117).[14] As with other "postpeasant" contexts (Barnes 1954), despite the clearly capitalist nature of the English rural economy, we still see some distinctly peasant-like attitudes in certain twentieth-century English narratives about community and the landscape. We see some of the economic practices of the peasantry in contemporary strategies adopted within the rural sector – from the way in which the "balance" invoked by S was centred around her extended family, working closely with their base of land and animals, to how large estates like that owned by R support themselves through sending out children to work in the wage economy, a practice that is also common for peasant households (Wolf 1966, 67). Returning to the work of David Matless (1988), *Landscape and Englishness* describes competing visions of what the English landscape that emerged in the mid-twentieth century should be like. Matless recounts how, in contrast to the dominant, modernist vision of the English countryside as a space for recreation, "an organic relationship to land is presented as dependent on and necessary for an organic social order [and how] organicism envisaged an organic English body at odds with the planner-preservationist ideal of modern citizenship" (ibid., 32). For organicist thinkers like Lionel Picton, Lord Northbourne, and Harold John Massingham, maintenance of the soil was linked to maintenance of the social order: "Soil, family and community are to be nurtured together" (ibid., 227). Agricultural practices like composting were viewed as having pro-social and pro-ecological properties; clothes and tools should be designed so that they can rot away, being returned to the fertility of the earth through the "rule of return" (ibid., 154). And crucially, work was seen as a having key normative role – "Beauty in organic England would emerge unselfconsciously through labour" (ibid., 151). Although, as Matless points out, this organicist thread to English cultural imaginaries did not achieve hegemonic status, it nonetheless remains influential – having prompted the formation of major UK NGOs, such as the Soil Association (ibid., 152).

There are clear parallels between the organicism described by Matless in *Landscape* and the sort of ecologically groundedness of Broadland's norms, exemplified by Moss's "people's wetland" and particularly by Ewans' "Marshland economy". The peasants and marshmen of Broadland are recruited to defend a sort of "Broadland Organicism" – where modern tourism is deemed to destroy the wetland that the "Marshland economy" continually restored. I would go further, and suggest that not only does organicism reveal cultural attitudes in the Broads about the idealised unity of people and the land, realised through work, but that it shares many similarities with the ideologies that empower peasant political movements elsewhere in the world. As Wolf points out:

> simplified movements of protest among a peasantry frequently center upon a myth of a social order more just and egalitarian than the hierarchical present. Such myths may look backwards, to the re-creation

> of a golden age of justice and equality in the past, or forward, to the establishment of a new order on earth, a complete and revolutionary change from existing conditions.
>
> (Wolf 1966, 106)

Wolf's remarks, here, could just as easily be applied to the organicist ideology of Broadland – whether or not we're talking about Ewans' "Broadland Economy" in the past, or Moss's "working landscape" of the future – in which the commons was managed by the entire local community, with more hierarchical elements (such as the position of the manorial lord) being downplayed. The ideals of "good farming" as described by my contributors also reflect this pattern, although clearly they draw the lines of good and bad conduct somewhat differently to intellectuals like Moss and Ewans. Following Winkler (2005) and Morphy (2003), I'd suggest "good farming" constitutes an "aesthetics of proximity" "in which local people, through their direct connection with working the landscape, develop aesthetic appreciations different from 'bourgeois' distant viewers – viewing the 'beauty of the work' rather than the 'beauty of the land'" (Burton 2012, 53),[15] that socialises the senses of farmers in such a way that they can "get their eye in" to agricultural land (Morphy 2003, 258–259). Despite framing its aesthetic value differently, however, both Broadland's farmers and organicist intellectuals – like Redfield's peasants – agree that that the land has aesthetic value that carries normative force.

But this treatment of Broadland as a "postpeasant" culture is not without problems. Early economists, such as Alexander Chayanov, argued that:

> The first fundamental characteristic of the [peasant] farm economy of the peasant is that it is a family economy. Its whole organization is determined by the size and composition of the peasant family and by the coordination of its consumptive demands with the number of its working hands.
>
> (Chayanov 1931)

Max Weber, in his analysis of English economic history, also emphasised the role of the household in the medieval economy, arguing that the disconnection of rural enterprise from the household was a key feature of the development of capitalism and the common law legal system (Weber 1978, 149, 980). Both Weber and Chayanov identify a close connection between the household as a basic means of production and the peasantry; the domestic mode of production is cast as diametrically opposed to the capitalist mode of production, with enclosure facilitating the transition from the former to the latter.

This line has been followed by many anthropologists, particularly Teodor Shanin and Harvey Franklin, who see the household as the defining feature of peasant life. Shanin (1971, 30) characterises peasant households as the "basic nuclei of peasant society", each of which are:

> operated as a highly cohesive unit of social organization, with basic divisions of labour, authority and prestige on ascribed family lines. Generally, the head of the household was the father of the family or the oldest kin-member. His authority over the other members and over household affairs by peasant custom implied both autocratic rights and extensive duties of care and protection. The household was the basic unit of production, consumption, property holding, socialization, sociability, moral support and mutual economic help. Both the social prestige and the self-esteem of a peasant were defined by the household he belonged to, and his position in it, as were his loyalties and self-identification.
>
> (Ibid., 31)

This domestic focus ensures that the role of the market is necessarily circumscribed, and that all other forms of social relations are subordinate to those of the kin group (Thomas and Znaniecki 1971, 23; Sahlins 2003, 92–94). The general point made by Chayanov, and his followers in anthropology, is that the central focus of work in peasant societies is the household occupied by an extended family; this household–kin-group unit also determines wider social norms and ecological relations. Of course, if this is the case, then the presence of "peasant-like" structures in the individualised, profoundly market-oriented society of contemporary Broadland will need an alternative explanation.

This view of peasant societies as domestically oriented is applied to English history by Alan Macfarlane (1978).[16] Utilising documentary evidence from two English parishes, Macfarlane argues that when one searches for evidence of the domestic mode of production in medieval England, it is either absent or contested with frequent exceptions – from nuclear families to a thriving land market – exceptions so frequent that he argues that English society was individualistically, rather than domestically oriented, as far back as we have documentary evidence (ibid., 201). In later work, he argues that the real origin of English individualism is a bilateral kinship system (Macfarlane 1992, 180–185). He warns that:

> The attraction of the "from peasant to industrial" theory lies very deep in our hearts, since it also appeals to the still strong nineteenth century evolutionary mode of thought, with its idea of gradual growth from small, closed, immobile, technologically simple, subsistence economies where life was "nasty, brutish and short", towards the human, mobile, affluent society of modern western Europe and North America. It is, furthermore, attractive to think of this "progress" from "lower" to "higher" as a more-or-less continuous line.
>
> (Macfarlane 1978, 192)

In short, Macfarlane argues that the domestic mode of production was never a structurally dominant feature of English society, and as such, the English never had a true peasantry. If this is so, then Broadland's inhabitants cannot be "postpeasants".

Macfarlane's argument was controversial among Marxist historians – it destabilises their account of the Enclosure Movement – but his claim about the cultural depth and specificity of English individualism struck a chord with the Thatcherite spirit of the late 1980s (White and Vann 1983; Ryan 1988; Gamble 1993). But the validity of the historical argument Macfarlane makes is not at issue in the present discussion. Instead, there are two points about his argument that are relevant here. Macfarlane is right to be wary of grand historical narratives, especially seductive ones. I would add that the moral weighting Macfarlane describes above can be reversed – it is equally tempting to condemn the modern, capitalist world for having shattered "traditional" peasant society, imagined as both ecologically sustainable and socially cohesive. Indeed, this is precisely what the intellectual narratives of Broadland organicism do, to varying degrees. In short, Macfarlane's rejection of the conventional narrative about English rural life reveals something highly significant *about* that life today, irrespective of the historical situation. It shows us that self-interested individualism must be a key part of any model of Broadland society we advance; not only does it highlight our contemporary preoccupation with the individual, but it demonstrates that the historical evidence can be read as evidence for the existence of individualism far back in English history.

Macfarlane also reinforces the diversity of England's economic history, in a way that wards off any attempt to – in the manner of the early Broadland writers – depict the rural past in terms of the collapse of an isolated, static tradition in the face of modernity. In the Broads, tradition and modernity have a long and productive relationship – after all, one of the Broads' most treasured features – the drainage mills – are a sign of capitalist "improvement" par excellence (Williamson 1997, 165). Nor were the people of this region ever set apart from the rest of English society: from the very earliest periods of settlement, Broadland has been enmeshed in translocal legal and economic networks, its development guided as much by flows of resources, money, ideology, and opportunity, as by the passage of its rivers. If London had not had much appetite for beef supplied by the Highlands of Scotland, then the "isolated" marshmen would have been out of a job. If ladies in Norwich had not wanted feathers for their hats, then the lifestyle of the "feral" Breydon punt gunners would have been impossible.

An alternative means of conceptualising peasant societies exists, however, where the domestic mode of production is not so central, and instead the *entire village* is the basic unit of production, within which common land and resources play a key role. When discussing the core features of the English peasantry, historian R. H. Hilton identifies (as the third trait) "(3) They are normally associated in *larger units than the family* – that is, in villages or hamlets with greater or lesser elements of common property and collective rights according to the character of the economy" (Hilton 1975, 13, emphasis mine).[17] Stirling describes how Turkish villages own common land – distinct from both private and public property – that is assiduously guarded by the entire village community from interlopers from neighbouring villages (Stirling 1971, 40–41). In Egypt, the prominence of the village over the kin group in peasant life is even more pronounced, as "the village or its quarter, not the house, makes up the entity, a community more important in many ways than the family or clan" (Ayrout 1971, 49). A particularly clear case of commoning as a key component of a peasant economic system is the Russian *mir*, or rural commune (Dunn and Dunn 1967, 9–12). As Gamst summarises:

> The mir consisted of a group of households, each of which had the right by virtue of its membership to hold plots of land in various categories (crop land, pasture, hay field, forest lot, and so on). Tenure was vested in the mir as a whole, however, and the land was subject to periodic redistribution among the constituent households, usually adult males. The assembly was known as the skhod, from the verb "skhodit", meaning to come together. Through its agent, the starosta (literally "elder"), the skhod conducted all transactions with individuals (or other social units) and with the state on behalf of its members.
>
> (Gamst 1974, 36)

Even more suggestive of the English situation where "the common" symbolises powerful imaginaries, is the fact that the word *mir* itself also means "peace" and "world". The *mir*, like the common, is a case of collective management relations distributed across every individual (or, at least, heads of households) in a village, which also has both normative and geographical connotations. Importantly, it is not the house, but the village, that is fundamental to the *mir*. The social representation of a village as a fundamental unit of "solidarity" is influential, being an important feature of administrative discourse in rural Laos (High 2006, 26–32) – where "appeals to 'working together', 'the common good' and ... (solidarity) evoke ideals of what a village could and should be" (ibid., 30–31) High goes on to describe how theorisations of Asian villages among social scientists vary between the construction of the village as a traditional, primordial social unit (e.g. Ireson 1996, 219) with its own "moral economy" (Scott 1976) and scholarship arguing that the village is a conceptual and administrative instrument of state power (e.g. Breman 1988; Kemp 1988). High summarises in the following way:

> While we may accept that the village is a rather recent, state-sponsored invention, it does not follow that all contemporary experiences of village are purely administrative experiences.... In contemporary Laos, the concept of village is bound up with concepts of belonging, place, mutual support, and aspiration. And these ideas are amorphous. They are never singular, and are often contradictory.
>
> (High 2006, 36)

The contemporary power of rural locality – regardless of the historical character – finds expression in numerous other ethnographic cases, such as in how established farmers and new entrants into farming in Tasmania both draw upon discourses of place-based locality – rather than domesticity – to construct their own narratives of dwelling and valorous work (Smith 2016). Fillaili, further, points out that localocentric solidarity – as opposed to government intervention or familial collaboration – provides the basis for resilience in the face of regular flooding in Kabupaten Sragen, Java (Fillaili 2016). The complexity of the working experience of local landscapes emphasised by High is also reflected in Anna Tsing's vivid descriptions of peasant landscapes and the vernacular knowledge that arises from them. Tsing points out that, in Yunnan, "Peasant forests were a modern object – a result of decentralisation – not an old one, and the goal of forest experts was to make modern rationality possible" (Tsing 2015, 187), while in Japan they are a subject of nostalgic attachment to the economic past, focused upon *iriai* common rights to the woodland, shared by local villagers (Tsing 2015, 180–187).

Without wishing to speculate about the precise nature of Broadland's medieval economy, this alternative model of peasant societies helps explain how "peasant-like" cultural forms – namely the association of working the land with normative social conduct – remain so influential in Broadland society today. If it is the commons and its parish – and not the household and its family – that is acting as the focus for these ideas, then the absence of the domestic mode of production would not diminish their power. So long as parochial communities continue to exist, then so will these attitudes. To be clear, I am not advancing an alternative to the domestic mode of production – say, *the parochial, or manorial mode of production* – and claiming that this holds in the Broads today, or that it ever did historically.[18] Rather, I suggest that the common and the parish simply act as an alternative way of "making sense" of space and human settlement, as opposed to the familial household or the individual business. The presence of individualism, a wage economy, and private land ownership are not fundamentally incompatible with the common, in the same way as these structures are said to be with the domestic mode of production by the scholars already cited; as such, it is possible for common norms and individualistic, capitalist principles to be held in balance, *and not at arm's length.* As Walker states, "The common is not to be confused with public property. It is a collective, productive resource that is antithetical to property, whether public or private" (Walker 2015). Ellis concurs, observing that: "Common property resources are rarely based on ownership in its legal sense" (Ellis 1993, 266). Rather than the commons existing as an alternative form of property rights to private and state property, the commons exists as a separate category of economic life entirely. Although privatisation may interfere with the realisation of this category in various ways (as we shall see in Chapter 4), it is *not* because it constitutes an opposing form of ownership. As such, what truly limits the commons, I suggest, is not the mere presence of wage labour, private property, or the market, but rather the related processes of the division of labour, and enclosure. It is these latter processes that shape the daily interactions of Broadland residents, moving people off the land, and thus reducing the role of the commons to a symbolic one, rather than that of a structure of day-to-day practice.

To better understand this process, it is helpful to consider the work of Pierre Bourdieu, and how it has been applied so far to the study of farmers' attitudes towards work and the landscape. Burton utilises the concept of habitus to explore the aesthetic of tidiness and its association with "good farming", pointing out that:

> For farmers, landscapes do not simply reflect established or historical aesthetic preferences. Rather, as landscapes play an important role in transferring flows of cultural and social capital between individuals and generations, the cultural meaning of being a farmer is heavily embedded in the landscape itself
>
> (Burton 2012, 66),

arising through the lifelong socialisation of children on family farms. As Riley points out, this means that many farmers think of farmed space "relationally", such that

> the material space of the farm becomes indivisible from those people who have historically managed and created that landscape, and uncovers the crucial issue of how *past practice develops into a moral framework that guides and constrains how practices are performed today.*
>
> (Riley 2008, 1285, emphasis mine)

As progressively fewer people participate in that work, however, fewer acquire the habitus connected with farming, and those norms begin to break down. Some 90 per cent of the Norfolk farmers surveyed by Hall and Pretty disagreed with the claim that people would never take advantage of or abuse their land (Hall and Pretty 2008a, 2), indicating a perceived lack of shared norms between farmers and non-farmers. Crucially, agricultural knowledge is cited as a key factor in building trust between government agents and farmers (Hall and Pretty 2008b, 398–399) – it is arguable that the same applies, to a lesser extent, to everyone with whom a farmer comes into contact. But the fact that such a large proportion of the general public in England no longer have the requisite knowledge to participate in the agricultural common landscape, has left the agricultural sector somewhat socially isolated, and marginalised in public discourse (Talbot and Walker 2007; Hall 2008). Compared to such issues as welfare, immigration, and healthcare, Britain's urban and suburban populations are largely ignorant or apathetic about the needs and norms of farming. To follow Hall in utilising Bourdieu's language, the symbolic capital of the agricultural field no longer translates effectively into social capital. In communities where a majority of the population work in rural industries, such as the Norwegian fishing village studied by J. A. Barnes,

> people living and working together inevitably have conflicting interests but in general they have also a common interest in the maintenance of existing social relations. Individual goals must be attained through socially approved processes, and as far as possible the illusion must be maintained that each individual is acting only in the best interests of the community.
>
> (Barnes 1954, 50)

But in among the commuters, tourists, second-home owners, and recent arrivals of Broadland's dormitory villages, such collective interests simply don't exist, leaving many farmers feeling isolated and misunderstood, their common sense no longer held in common with those who share their surroundings.

What is lacking in this situation is a "structuring structure" according to which English social relations tend to develop; namely, the habitus of commons. Bourdieu explains that:

> The structures constitutive of a particular type of environment (e.g. the material conditions of existence characteristic of a class condition) produce habitus, systems of durable, transposable dispositions, structured structures predisposed to function as structuring structures, that is, as principles of the generation and structuring of practices and representations which can be objectively "regulated" and "regular" without in any way being the product of obedience to rules ... collectively orchestrated without being the product of the orchestrating action of a conductor.
>
> (Bourdieu 1977, 72)

A classic example of Bourdieu's own work on how class-dictated social space can engender a habitus peculiar to it, is his description of the Kabyle household:

> The meaning objectified in things or places is fully revealed only in the practices structured according to the same schemes which are organized in relation to them (and vice versa).... The house is organized according to a set of homologous oppositions – fire:water :: cooked:raw :: high:low :: light:shade :: day:night :: male:female :: *nif: hurma* :: fertilizing:able to be fertilized. But in fact the same oppositions are established between the house as a whole and the rest of the universe, that is, the male world, the place of assembly, the fields, and the market.... But one or the other of the two systems of oppositions which define the house, either in its internal organization or in its relationship with the external world, is brought to the foreground, depending on whether the house is considered from the male point of view or the female point of view: whereas

> for the man, the house is not so much a place he enters as a place he comes out of, movement inwards properly befits the woman.
>
> (Bourdieu 1977, 90–91)

The important distinction between Bourdieu's formulation and the kind of structuralist anthropology on which it drew was described as follows by Lane:

> The locus of meaning had thus shifted from the system of binary oppositions laid bare by the detached intellectual observer to the movements and perspectives of the Kabyles themselves. *Meanings were generated not at the level of a purely theoretical, disembodied structure of difference but by the Kabyles in their everyday actions and movements*, in their 'practice' as Bourdieu was to put it....
>
> (Lane 2000, 98, emphasis mine)

In Bourdieu's own words:

> All the actions performed in a space constructed in this way are immediately qualified symbolically and function as so many structural exercises through which is built up practical mastery of the fundamental schemes, which organize magical practices and representations: going in and coming out, filling and emptying, opening and shutting, going leftwards and going rightwards, going westwards and going eastwards, etc. Through the magic of a world of objects ... each practice comes to be invested with an objective meaning, a meaning with which practices – and particularly rites – have to reckon at all times, whether to evoke or revoke it.
>
> (Bourdieu 1977 op. cit.)

The key difference between Bourdieu and a structuralist account of the same processes is the view that structures are not just patterns of thought but patterns of action – that is, *habitus*. My suggestion is that what *the household does for the Kabyle (a peasant society) the common once did for the English – and does still, to a limited extent, for farmers and other land managers*. Just as the broad scheme of social relations between genders, between residents and strangers, and between animals and humans is redolent in the layout of physical space in the *Kabyle* household, so the shape of the household determines certain patterns of movement and work that reinforce that scheme in the minds and bodies of the Kabyle. Farmers continually reinforce the habitus of commoning, by embodying particular common, normative practices in the labours they perform. But what power, beyond the farming community, and that of other land managers, does the common command?

In Ortner's discussion of the anthropological study of what she terms "key symbols", she identifies one particular type of key symbol that highlights the kind of role commons seem to have in wider English society: "Elaborating symbols [work by] providing vehicles for sorting out complex and undifferentiated feelings and ideas, making them comprehensible to oneself, communicable to others, and translatable into orderly action" (Ortner 1973, 1340). This is precisely what we saw happening at the beginning of the chapter, I suggest, when L used productive work to create "common ground". His use of physical action and phraseology invokes the common as a key symbol, for the purpose of elaborating the proper relationship between people, place, and knowledge. Indeed, I would go as far to say that the common – or rather, the parish, with a common land, commoners, and common sense – acts as a "root metaphor" (ibid., 1341) in English culture (see Figure 3.3). As Strang points out regarding rural life in Dorset, "Although interactions with the environment have changed radically, people still draw upon images of past communities and systems of management to formulate ideas about the present and construct 'ideal' models" (Strang 2004, 10). The common, I suggest, is one such image of past communities and systems of management. This metaphor is actively deployed through how one acts in the landscape, serving as the analogical basis for modern institutions of property ownership and economic propriety (Douglas 1987). Farmers, for example, plan, plough, plant, fertilize, spray, and harvest – leaving direct physical traces in the landscape. The recreation of these physical traces reinforces embodied standards of "good farming" according to which they were first created. Although their land is privately owned, it exists within a shared community of fellow farmers, where each farm physically manifests the habitus of each farmer, and in turn creates the conditions for that habitus' reinforcement. What for today's farmers and farm workers is performed

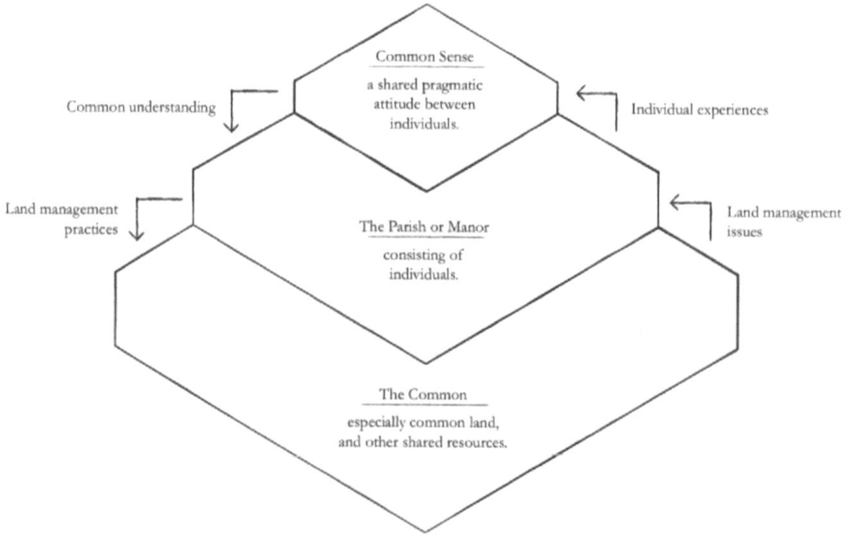

Figure 3.3 The common as a root metaphor.

through everyday practice, was once an everyday practice too for the majority of the English population, engendering a habitus of the common across society. Contemporary attitudes towards shared land, labour, "community", and resources, are a root metaphor, I suggest, grounded and naturalised in the premises of this common experience (Morphy 2011, 262).

Enclosure, therefore, occurs not just at a legal or economic level, but at the level of habitus. Taking the example of the planting of hedges as fields were enclosed, Blomley argues:

> The hedge both helped to concretise a new set of controversial discourses around land and property rights, and aimed to prevent the forms of physical movement associated with the commoning economy.... [T]his spatial discipline was socially directive: in other words, it drew from and helped produce an emergent set of social hierarchies that rested on developing conceptions of private property. I wish to point to the body [of the commoner and their beasts, as the] site upon which new forms of discipline, materialised in the hedge, were to be realised.
>
> (Blomley 2007, 5)

Enclosure entails a restrictive habitus, where access to the countryside is physically prevented, and sensitivity to highly exclusive categories of private property becomes paramount. Commoning, by the same token, can produce a very different kind of habitus, still found among land managers. The ongoing process of enclosure restricts the breadth of the habitus of the common, and so the common only attains broader relevance in English society as a root metaphor, rather than as a habitus. Much reduced though they are, commons – like Mousehold Heath in Norwich – represent crucial sites where this root metaphor can still be internalised through practice by the wider population, but to a limited extent.[19] Recent academic scholarship on the commons has underscored this view – also seen in historical narratives about common land as a site of community organising and management – that the commons, many of which have been reduced to halcyon-tinted collective memory, are founded upon deep-seated attitudes and beliefs. Although these attitudes are often unacknowledged or undermined by more dominant, pro-market or pro-state economic regimes, they nonetheless are deployed to sustain public life and social relations (Rose 1994; Blomley 2008, 319; Amin and Howell 2016). By depriving the English people of their common premises – that is, large areas of land that are managed and utilised collectively – common premises of the intellectual sort are lost too. Understanding how engagement with the landscape through practice takes place will be developed further in the next chapter.

Conclusion: the institution of common ground

At the beginning of this chapter, I reflected upon the controversies around the management of Mousehold Heath, a fragment of common land in the heart of Norwich. L's position – that work of particular kinds creates a connection between people with diverse interests – is founded upon a very specific set of assumptions, assumptions that, I argue, are encoded in the root metaphor of the common. Inspired by Veronica Strang's analysis of the meaning of water (2004), and drawing upon Pierre Bourdieu's analysis of the Kabyle house (1977), Sherry Ortner's understand of key symbols (1973), and the geographer David Matless's analysis of early twentieth-century organicism (1998), I have suggested that the common has *doxic* effects within English society – as a habitus of farmers and other land managers, and as a root metaphor for everyone else. The creation or imagination of common places that can only be managed collectively incorporates, objectifies, and reinforces the particular attitude identified by L. Commoning (through making) and enclosure are therefore two *competing* forms of habitus that reinforce the commons, and individualism of a particular kind – self-interested, *possessive* individualism[20] – respectively. With reference to classic anthropological discussions of English individualism (Macfarlane 1978, 1992; Dumont 1986), I suggest that the commons and possessive individualism both provide a framework through which social relations between individuals (the individual also being a *doxic* feature) can be created and sustained. The power of private ownership and market exchange at the expense of common rights, means that it is possessive individualism that is in a hegemonic position in the Broads today. The common still has a role to play as a root metaphor – symbolising moral behaviour, good sense, and understanding culled from experience – but it is only experienced through the *habitus* of commoning under certain circumstances, such as by land managers.

What we also see from Broadland's environmental histories and futures is how peasant-like qualities can exist in societies even where the domestic mode of production is absent. Indeed, rather than treat peasantry as a Weberian ideal type, it might be more useful to treat peasant societies in terms of a Wittgensteinian family resemblance (Ahmed 2010, 66–87). No single peasant society need express all the features associated with peasantry, for all peasant societies to resemble one another to a degree. Indeed, as Rapport points out, this same pattern reflects perceptions of community and place in rural England:

> Perceptions in Wanet … form a family of resemblances, a bundle of partially overlapping cognitive constructions, and, as we shall see, it is individual interpretation of the relations of the moment which determines which consideration is pertinent, which construction is salient, when.
>
> (Rapport 1993, 51)

We need not follow Weber and Chayanov in insisting that all peasant like qualities spring from the domestic mode of production, when the ethnographic material indicates otherwise. Fruitful comparison can be struck by, as Shanin exhorts us to do, treating peasantry as a process, rather than a static condition. Consider the following description of peasant sociality by Wolf:

> *The peasant stands, as it were, at the center of a series of concentric circles, each circle marked by specialists with whom he shares less and less experience, with whom he entertains fewer and fewer common understandings....* There are those close to him, peasants like himself, whose motives and interests he shares and understands, even when his relations with them are wholly tangential.... *These do not form a group characterised by enduring social relationships, but a category of people with whom interaction and understandings are possible on the basis of common premises....* Characteristically, however, there is a shift of attitudes when the peasant confronts the person who has a lien on his surplus of rent or on his surplus of profit: the merchant, the tax-collector, the manager of a putting-out system who farms out craft production to the villages ... the labor contractor.... *Economic interests are directly opposed, and are not counter-balanced by more personal involvements. Thus social distance is reinforced by an absence of shared experience.* Hence, where we find peasants involved in network markets, we also find that the merchant or storekeeper – even when he resides in the village – continues to be regarded as a stranger and outsider.
>
> (Wolf 1966, 47, emphasis mine)

Although they are, squarely, "postpeasants", Broadland's people take an attitude to social life that reflects the dynamic expressed by Wolf above. The common is situated in the context they locate themselves – in the "farming community" say – while those who do not share that same working landscape – that same taskscape – and the common sense that comes with it, are placed at greater and greater degrees of social remove. Where they differ, however, is in their attitude towards the market, and the profit motive. But just because self-interest is deemed to be natural in the heavily marketised economy of Broadland, the power of the common as a doxic, root metaphor shouldn't be underestimated – as Burton points out, there has been little change to the productivist mentality of European farmers, despite extensive government intervention, with significant impacts upon biodiversity and water quality (Burton, Kuczera, and Schwarz 2008, 16).

* * *

Common sense reflects how English people think about both resources and social relations through a single frame – a move that has its roots in an imagined parochial ground, through both academic and popular discussions about the past. Common land serves as common ground – a place of collective interest and labour with which it is also good to think, that yields good ideas, as well as useful raw materials and products. In *Conversations in Colombia*, Stephen Gudeman and Alberto Rivera suggest that European folk culture represented a wellspring of ideas that informed later conversations by both political economists and cultures (1990). Here, I'd like to put forward "the common" and "common sense" as an example of such concepts, which are sustained by a dialogue between scholars and diverse publics. As indicated by wood carving on Mousehold Heath, practical work and reason coincide in English lived experience. Though this can manifest in any shared material context – something as simple as a piece of wood being carved into a usable object – the most powerful site for the expression of this common sense, propelling what is a root metaphor to the position of a structuring structure, is common land. The fact that individualistic self-interest is prevalent within English culture and a key part of economic life does not preclude the possibility of common concern – as Strathern points out, to try to disentangle self-interest from other-interest is nonsense (Strathern 2016, 168–171). The two exist in tandem, and together structure English practices in social space.

Notes

1 This distribution was typical of the conservation groups with whom I volunteered, though the distribution between these groups would vary according to factors like location, the timing of task days, and facilities. Because TCV Norwich was based in an urban area, its Environmental Action Group attracted far more unemployed people, whereas the Bure Valley Living Landscape group – based in the relatively affluent village of Acle – attracted more retired people. Strumpshaw Fen – which provided accommodation for long-term volunteers – attracted a lot of students.

2 See Chapter 4 for a more detailed description of succession.

3 A now infamous quote from former Prime Minister, Margaret Thatcher, from an interview she gave to Women's Own in 1987 –

> I think we have gone through a period when too many children and people have been given to understand "I have a problem, it is the Government's job to cope with it" … and so they are casting their problems upon society, and who is society? There are individual men and women and there are families, and no government can do anything except through people and people must look to themselves first. It is our duty to look after ourselves and then after our neighbour … and people have got the entitlements too much in mind without the obligations.… There is no such thing as society.
>
> (Moore 2010)

This claim was made in a context of widespread academic scepticism about the usefulness of "society" as a concept, as evinced by the 1989 debate between Marilyn Strathern, Christina Toren, John Peel, and Jonathan Spencer on the subject, which directly referred to Thatcher's remarks (Ingold 1996). This chapter is a testament to how controversial this denial of society was, and still is, in English society.

4 I deal with the shades of meaning attached to the absence of common sense in detail in Chapter 4.

5 Turbaries are excavations where peat is extracted for fuel, and constituted one form of common right traditionally enjoyed by commoners.

6 A particularly severe flood struck the region in 1287, when the sea broke into the Broads region – an event recorded by John of Oxnead of St Benet's Abbey, which sat by the side of the Bure (John of Oxnead 1859).

7 Staithe is a dialect word – common in the East of England, derived from the Old English stæð – for a small quayside on a river or channel at which boat traffic may dock. Historically these were used in the Broads for commercial purposes, but are nowadays more often used by pleasure craft.

8 As Moss (2001) has argued, the increases in nitrogen (from agricultural waste) and phosphorus (from treated sewage) did not by themselves lead to the loss of the Broads' water clarity – indeed, in the early twentieth century, these nutrients caused the existing populations of water plants to grow rather better. Rather, elevated concentrations of these nutrients simply made it more likely that eutrophication would result. Once the eutrophicated state was triggered – which, Moss contends, was a result of the sudden collapse in the population of zooplankton that predate upon algae – it has been extremely difficult to reverse.

9 The Nature Conservancy Council and the Countryside Commission were amalgamated into Natural England in 2006. Like its predecessors, Natural England is a non-departmental public body that is responsible for the preservation and improvement of England's natural environment.

10 Although O treated these positions as simple observations, obvious to anyone who works with other people, they both have a life as particular, and entirely arguable, theoretical positions within social science. Both Marxist and Hobbesian views of human nature here have entered into the popular discourse of English society, to the point that people view both as simple statements of fact – often, as O demonstrates, the same people.

11 Although in the minimal sense this includes other farmers, farm workers, and landowners, it could also be expanded to included long-term or knowledgeable "locals" who were involved in agriculture indirectly.

12 Brian's grandmother Phoebe was descended from the Howard, Kerrison, and Mutton families – Howard Level, Kerrison Level, and Mutton Level are all marshland areas in the Broads, each of which was drained by a single mill, tended by the family after which the level and its mill were named.

13 "Social housing" is a class of housing stock, owned by councils or not-for-profit housing associations, that is more affordable than private ownership or renting.

14 Harvey uses "common sense" in the classic, Gramscian sense, rather than in the sense of the emic definition I developed in Chapter 1 (2007, 39).

15 To Burton's point, I would add that "beauty" for farmers and bourgeois viewers can be expressed and experienced quite differently. While bourgeois aesthetics involves a rapturous "quiet enjoyment" of natural beauty, aesthetic appreciation for rural workers – as we have seen – is more expressed and articulated as a sort of "quiet satisfaction" of a job well done.

16 Macfarlane draws heavily on Shanin's work, and therefore can be located within the squarely Chayanovian tradition.

17 In their review of Macfarlane's *Origins*, White and Vann allege that Macfarlane overlooked this qualification in Hilton's work on the English peasantry (White and Vann 1983, 350).

18 If we follow Macfarlane's reasoning, it is arguable that reimagining community as a parochial commons (rather than a network of semi-autonomous extended families) was another outcome of the English bilateral kinship system.

19 An example of which would be the practice of "making things", as described by L at the beginning of this chapter.

20 I discuss possessive individualism in greater detail in Chapter 5.

References

Adams, Tim. 2011. "Who Owns Our Green and Pleasant Land?" *Guardian*, August 7, 2011, sec. UK news. www.theguardian.com/lifeandstyle/2011/aug/07/tim-adams-who-owns-britain.

Ahmed, Arif. 2010. *Wittgenstein's "Philosophical Investigations": A Critical Guide*. Cambridge: Cambridge University Press.

Albright, Karen, Parama Chaudhury, Alvis Dunn, Pushan Dutt, Vibha Gaba, Saran Ghatak, Brunson Hoole, Taylor Sisk, Triadafilos Triadafilopoulos, and Guobin Yang. 2002. *Dictionary of the Social Sciences*, edited by Craig Calhoun. New York, NY: Oxford University Press.

Allen, Robert C. 1992. *Enclosure and the Yeoman: The Agricultural Development of the South Midlands 1450–1850*. Oxford; New York, NY: Clarendon Press.

Allmer, Thomas. 2010. "Summary of Commonwealth by Hardt and Negri". University of Salzberg. www.allmer.uti.at/wp-content/uploads/2011/11/Summary Commonwealth_ThomasAllmer1.pdf.

Amin, Ash, and Philip Howell. 2016. "Thinking the Commons". In *Releasing the Commons: Rethinking the Futures of the Commons*, edited by Ash Amin and Philip Howell, 1–17. London; New York, NY: Routledge.

Anderson, Esther. 2016. "Meeting the Locals: Working Holidaymakers' Experiences of Community in a Regional Australian Agricultural Town". In *Rural Dwelling and Resilience*. University of Sydney. www.academia.edu/30567766/Meeting_the_locals_Working_holidaymakers_experiences_of_community_in_a_regional_Australian_agricultural_town.

Austin, Janice, Jayne Greenacre, and Brian Grint. 2011. *The Search for Eugenia Fynch: The Story of Norfolk's Unknown Victorian Photographers*. The Acle Community Archive Group.

Ayrout, Henry Habib. 1971. "The Village and the Peasant Group". In *Peasants and Peasant Societies*, edited by Teodor Shanin, 30–35. London: Penguin Books.

Bacon, K. P. 1993. "Enclosure in East Norfolk: Particularly Non-Parliamentary Encosure and Consolidation in the Hundreds of Happing and Flegg". Unpublished MA Thesis, Centre of East Anglian Studies: University of East Anglia.

Barnes, J. A. 1954. "Class and Committees in a Norwegian Island Parish". *Human Relations* 7 (1): 39–58. https://doi.org/10.1177/001872675400700102.

Blomley, Nicholas. 2007. "Making Private Property: Enclosure, Common Right and the Work of Hedges". *Rural History* 18 (1): 1–21. https://doi.org/10.1017/S0956793306001993.

Blomley, Nicholas. 2008. "Enclosure, Common Right and the Property of the Poor". *Social & Legal Studies* 17 (3): 311–331. https://doi.org/10.1177/0964663908093966.

Bourdieu, Pierre. 1977. *Outline of a Theory of Practice:* Translated by Richard Nice. Cambridge Studies in Social and Cultural Anthropology. Cambridge University Press. www.cambridge.org/core/books/outline-of-a-theory-of-practice/193A11572779B478F5BAA3E3028827D8.

Breman, Jan. 1988. *The Shattered Image : Construction and Deconstruction of the Village in Colonial Asia*. Dordrecht: Foris [for] Centre for Asian Studies, Amsterdam.

Broads Authority. 2013. "Strategic Priority Objectives, Projects and Key Milestones 2013/14". The Broads Authority.

Broads Authority. 2014. "History of the Broads Authority". UK Government. The Broads Authority. 2014. www.broads-authority.gov.uk/broads-authority/who-we-are/history-of-the-broads-authority.

Bronner, Simon J. 1984. "The Early Movements of Anthropology and Their Folkloristic Relationships". *Folklore* 95 (1): 57–73.

Burton, Rob J. F. 2012. "Understanding Farmers' Aesthetic Preference for Tidy Agricultural Landscapes: A Bourdieusian Perspective". *Landscape Research* 37 (1): 51–71. https://doi.org/10.1080/01426397.2011.559311.

Burton, Rob J. F. 2004. "Seeing Through the 'Good Farmer's' Eyes: Towards Developing an Understanding of the Social Symbolic Value of 'Productivist' Behaviour". *Sociologia Ruralis* 44 (2): 195–215. https://doi.org/10.1111/j.1467-9523.2004.00270.x.

Burton, Rob. J. F., Carmen Kuczera, and Gerald Schwarz. 2008. "Exploring Farmers' Cultural Resistance to Voluntary Agri-Environmental Schemes". *Sociologia Ruralis* 48 (1): 16–37. https://doi.org/10.1111/j.1467-9523.2008.00452.x.

Cahill, Kevin. 2002. *Who Owns Britain and Ireland*. Edinburgh: Canongate Books.

Chandler, Michael. 2012. *Robert Kett and the Norfolk Rebellion*. Nottingham; JMD Media.

Chayanov, A. V. 1931. "The Socio-Economic Nature of Peasant Farm Economy." In *A Systematic Source Book in Rural Sociology*, edited by Pitirim Aleksandrovich Sorokin, Carle Clark Zimmerman, and Charles Josiah Galpin, 144–145. Minneapolis: University of Minnesota Press.

Cocker, Mark. 2008. *Crow Country*. London: Vintage.

Dalton, George, H. Russell Bernhard, Beverly Chiv̄as, Beverly Chiñas, S. H. Franklin, David Kaplan, and Eric R. Wolf. 1972. "Peasantries in Anthropology and History [and Comments and Replies]". *Current Anthropology* 13 (3/4): 385–415.

Dobbs, Thomas L., and Jules N. Pretty. 2001. "The United Kingdom's Experience with Agri-Environmental Stewardship Schemes: Lessons and Issues for the United States and Europe". *University of Essex Centre for Environment and Society*, Occasional Paper no. 2001–1. http://agris.fao.org/agris-search/search.do?recordID=US2012207016.

Douglas, Mary. 1987. *How Institutions Think*. London: Routledge & Kegan Paul.

Dumont, L. 1986. *Essays on Individualism: Modern Ideology in Anthropological Persective*. Chicago: University of Chicago Press.

Dunn, Stephen Porter, and Ethel Dunn. 1967. *The Peasants of Central Russia*. New York, NY: Holt, Rinehart and Winston.

Durie, Taihakurei (Eddie). 2011. "Cultural Appropriation". In *Ownership and Appropriation*, edited by Veronica Strang and Mark Busse, English edition. A.S.A. Monographs. Oxford; New York, NY: Berg Publishers.

Dutt, William. 1903. *The Norfolk Broads*. London: Methuen & Co.

Ellis, Frank. 1993. *Peasant Economics: Farm Households in Agrarian Development*. Cambridge: Cambridge University Press.

Emerson, Peter Henry. 1885. *Life and Landscape on the Norfolk Broads*. Photographic Collection. The Metropolitan Museum of Art [The Met Museum]. www.metmuseum.org/art/collection/search/290484.

Emerson, Peter Henry. 1893. *On English Lagoons*. London: David Nutt.

Ewans, Martin. 1992. *The Battle for the Broads: A History of Environmental Degradation and Renewal*. Lavenham, SFK: Terence Dalton Ltd.

Fillaili, Rizki. 2016. "Flooding and Resilience in Indonesia". In *Rural Dwelling and Resilience*. Sydney: University of Sydney.

Frake, Charles. 1996. "A Church Too Far Near a Bridge Oddly Placed: The Cultural Construction of the Norfolk Countryside". In *Redefining Nature: Ecology, Culture and Domestication*, edited by R. F. Ellen and Katsuyoshi Fukui. Oxford: Berg.

Franklin, Harvey. 1971. "The Worker Peasant in Europe." In *Peasants and Peasant Societies*, edited by Teodor Shanin, 98–103. Harmondsworth: Penguin Books.

Gamble, Andrew. 1993. "The Entrails of Thatcherism". *New Left Review* 1 (198): 117–128.

Gamst, Frederick C. 1974. *Peasants in Complex Society*. Basic Anthropological Units. New York, NY: Holt, Rinehart, and Winston.

George, Martin. 1992. *Land Use, Ecology and Conservation of Broadland*. Chichester: Packard.

Glassman, Jim. 2006. "Primitive Accumulation, Accumulation by Dispossession, Accumulation by 'Extra-Economic' Means". *Progress in Human Geography* 30 (5): 608–625.

Grint, B. 1989. *An Acle Chronicle*. Cromer: Poppyland Publishing.

Grint, B. S. 1984. *The Halvergate Chronicles*. Privately printed.

Gudeman, Stephen, and Alberto Rivera. 1990. *Conversations in Colombia: The Domestic Economy in Life and Text*. Cambridge: Cambridge University Press. www.cambridge.org/core/books/conversations-in-colombia/83B14D4A41AC74F161D443D7E930A7AE.

Hackett, Michael, and Alan Lawrence. 2014. "Multifunctional Role of Field Margins in Arable Farming". Environmental Assessment CEA.1118. Cambridge Environmental Assessments – ADAS UK Ltd. Cambridge: European Crop Protection Association. www.ecpa.eu/sites/default/files/Field%20Margins%20Arable%20Farming_V02%20(1).pdf.

Hall, Jilly. 2008. "The Role of Social Capital in Farmers' Transitions Towards More Sustainable Land Management". PhD Dissertation. Colchester: University of Essex.

Hall, Jilly, and Jules Pretty. 2008a. "Unwritten, Unspoken: How Norms of Rural Conduct Affect Public Good Provision". Paper presented in *Rural Futures: Dreams, Dilemmas, Dangers*, Compilation Conference, University of Plymouth, UK, 1–4 April.

Hall, Jilly, and Jules Pretty. 2008b. "Then and Now: Norfolk Farmers' Changing Relationships and Linkages with Government Agencies During Transformations in Land Management". *Journal of Farm Management* 13 (6): 393–418.

Hardt, Michael, and Antonio Negri. 2009. *Commonwealth*. Cambridge, MA: Harvard University Press.

Harvey, David. 2007. *A Brief History of Neoliberalism*. Oxford: Oxford University Press.

High, Holly. 2006. "'Join Together, Work Together, for the Common Good – Solidarity': Village Formation Processes in the Rural South of Laos". *SOJOURN: Journal of Social Issues in Southeast Asia* 21 (1): 22–45.

Hilton, Rodney Howard. 1975. *The English Peasantry in the Later Middle Ages: The Ford Lectures for 1973 and Related Studies*. Oxford: Clarendon Press. https://quod.lib.umich.edu/shcgi/t/text/pageviewer-idx?cc=acls;c=acls;idno=heb01202.0001.001;node=heb01202.0001.001%3A3;rgn=div1;view=image;page=root;seq=23.

Ingold, Tim. 1996. *Key Debates in Anthropology*. London: Routledge.

Ireson, W. Randall. 1996. "Invisible Walls: Village Identity and the Maintenance of Cooperation in Laos". *Journal of Southeast Asian Studies* 27 (2): 219–244. https://doi.org/10.1017/S0022463400021032.

Irvine, Richard D. G., Elsa Lee, Miranda Strubel, and Barbara Bodenhorn. 2016. "Exclusion and Reappropriation: Experiences of Contemporary Enclosure among Children in Three East Anglian Schools". *Environment and Planning D* 34 (5): 935–953. https://doi.org/10.17863/CAM.22.

Irvine, Richard, and Mina Gorji. 2013. "John Clare in the Anthropocene". *Cambridge Anthropology* 31 (1): 119–132.

Jeffrey, Alex, Colin McFarlane, and Alex Vasudevan. 2012. "Rethinking Enclosure: Space, Subjectivity and the Commons". *Antipode* 44 (4): 1247–1267. https://doi.org/10.1111/j.1467-8330.2011.00954.x.

John of Oxnead. 1859. *Chronica Johannis de Oxenedes*. London: Longman, Brown, Green, Longmans and Roberts.

Kaplonski, Chris. 2015. "Waiting for the Consumer to Catch Up: Sustainability and the Taste of Wine in Austria." Oxford Food Forum Annual Conference, 2 May.

Kemp, Jeremy H. 1988. *Seductive Mirage: The Search for the Village Community in Southeast Asia*. Dordrecht: Foris [for] Centre for Asian Studies, Amsterdam.

Lambert, J., J. N. Jennings, C. T. Smith, C. Green, and J. N. Hutchinson. 1960. *The Making of the Broads: A Reconsideration of Their Origin in the Light of New Evidence*. Royal Geographic Society Research Series 3. London: Royal Geographical Society and John Murray.

Lane, Jeremy F. 2000. *Pierre Bourdieu: A Critical Introduction*. London: Pluto Press.

Lankford, Bruce. 2016. "The Right Not to Be Excluded: Common Property and the Struggle to Stay Put". In *Releasing the Commons: Rethinking the Futures of the Commons*, edited by Ash Amin and Philip Howell, 89–106. London; New York, NY: Routledge.

Lewis, Oscar. 1951. *Life in a Mexican Village: Tepoztlán Restudied*. Champaign: University of Illinois Press.

Lobley, Matt, and Clive Potter. 2004. "Agricultural Change and Restructuring: Recent Evidence from a Survey of Agricultural Households in England" *Journal of Rural Studies* 20 (4): 499–510. https://doi.org/10.1016/j.jrurstud.2004.07.001.

Lobley, Matt, Clive Potter, Allan Butler, Ian Whitehead, and Ian Millard. 2005. "The Wider Social Impacts of Changes in the Structure of Agricultural Businesses". CRR Research Report No. 14: Final Report for Defra. Exeter: University of Exeter – Centre for Rural Research.

Lubbock, Richard. 1845. *Observations on the Fauna of Norfolk: And More Particularly on the District of the Broads*. Norwich: Charles Muskett.

Macfarlane, Alan. 1978. *The Origins of English Individualism: The Family, Property and Social Transition*. Oxford: Basil Blackwell.

Macfarlane, Alan. 1992. "On Individualism". *Proceedings of the British Academy* 82: 171–199.

Mcfarlane, Colin, and Renu Desai. 2016. "The Urban Metabolic Commons: Rights, Civil Society, and Subaltern Struggle". In *Releasing the Commons: Rethinking the Futures of the Commons*, edited by Ash Amin and Philip Howell, 145–160. London; New York, NY: Routledge.

Marx, Karl. 2000. *Karl Marx: Selected Writings*. Edited by David McLellan. 2nd edition. New York, NY: Oxford University Press.

Matless, David. 1998. *Landscape and Englishness*. 2nd edition. London: Reaktion Books.

Matless, David. 2014. *In the Nature of Landscape: Cultural Geography on the Norfolk Broads*. 1st edition. Chichester, West Sussex; Malden, MA: Wiley-Blackwell.

Miller, J., and H. Ayer. 1981. "Agrarian Justice: Thomas Paine's Social Security Program of 1797". *Educational Gerontology* 7 (4): 375–382.

Moore, Charles. 2010. "No Such Thing as Society: A Good Time to Ask What Margaret Thatcher Really Meant." *Daily Telegraph*, 27 September 2010, sec. Comment. www.telegraph.co.uk/comment/columnists/charlesmoore/8027552/No-Such-Thing-as-Society-a-good-time-to-ask-what-Margaret-Thatcher-really-meant.html.

Morphy, Howard. 2003. "For the Motion: Aesthetics Is a Cross-Cultural Category". In *Key Debates in Anthropology*, edited by Tim Ingold, 255–260. London: Routledge.

Morphy, Howard. 2011. "'Not Just Pretty Pictures' Relative Autonomy and the Articulations of Yolngu Art in Its Contexts". In *Ownership and Appropriation*, edited by Veronica Strang and Mark Busse, English edition. A.S.A. Monographs. Oxford; New York, NY: Berg Publishers.

Moss, Brian. 2001. *The Broads: The People's Wetland*. 1st edition. London: Collins.

Nonini, Donald M. 2006. "Reflections on Intellectual Commons". *Social Analysis: The International Journal of Social and Cultural Practice* 50 (3): 203–216.

Ortner, Sherry B. 1973. "On Key Symbols". *American Anthropologist* 75 (5): 1338–1346.

Patriquin, Larry. 2004. "The Agrarian Origins of the Industrial Revolution in England". *Review of Radical Political Economics* 36 (2): 196–216. https://doi.org/10.1177/0486613404264190.

Polanyi, Karl. 2001. *The Great Transformation: The Political and Economic Origins of Our Time*. 2nd Beacon Pbk. Boston, MA: Beacon.

Price, Patricia L. 2010. "Cultural Geography and the Stories We Tell Ourselves". *Cultural Geographies* 17 (2): 203–210.

Pryor, Francis. 2011. *The Making of the British Landscape: How We Have Transformed the Land, from Prehistory to Today*. London: Penguin.

Rackham, Oliver. 1997. *The History of the Countryside*. 2nd edition. History of the Countryside : The Classic History of Britain's Landscape, Flora and Fauna. London: Phoenix Giant.

Rapport, Nigel. 1993. *Diverse World-Views in an English Village*. Edinburgh: Edinburgh University Press.

Redfield, Robert. 1969. *Peasant Society and Culture: An Anthropological Approach to Civilization*. 5th edition. Chicago and London: University of Chicago Press.

Riley, Mark. 2008. "Experts in Their Fields: Farmer – Expert Knowledges and Environmentally Friendly Farming Practices". *Environment and Planning A* 40 (6): 1277–1293. https://doi.org/10.1068/a39253.

Robinson, John G., and Kent Hubbard Redford. 1991. *Neotropical Wildlife Use and Conservation*. Chicago: University of Chicago Press.

Rose, C. 1994. *Property and Persuasion: Essays on the History, Theory, and Rhetoric of Ownership*. Boulder: Westview.

Rosenman, E. 2015. "On Enclosure Acts and the Commons". BRANCH: Britain, Representation and Nineteenth-Century History. April.

Ryan, Alan. 1988. "English Individualism Revisited". *London Review of Books*, 21 January.

Sahlins, Marshall. 2003. *Stone Age Economics*. 2nd edition. London u.a.: Routledge.

Scott, James C. 1976. *The Moral Economy of the Peasant : Rebellion and Subsistence in Southeast Asia*. New Haven, CT: Yale University Press.

Shanin, Teodor. 1971. "A Russian Peasant Household at the Turn of the Century". In *Peasants and Peasant Societies*, edited by Teodor Shanin, 30–35. London: Penguin Books.

Smith, Jennifer. 2016. "New Farmers in Tasmania – Learning to Dwell". In *Rural Dwelling and Resilience*. Darlington: University of Sydney.

Stephenson, Carl. 1956. *Mediaeval Feudalism*. Ithaca, NY: Cornell University Press.

Stirling, Paul. 1971. "A Turkish Village". In *Peasants and Peasant Societies*, edited by Teodor Shanin, 30–35. London: Penguin Books.

Strang, Veronica. 1997. *Uncommon Ground: Cultural Landscapes and Environmental Values*. Oxford: Berg.

Strang, Veronica. 2004. *The Meaning of Water*. Oxford; New York, NY: Bloomsbury Academic.

Strathern, Marilyn. 2003. *Commons and Borderlands: Working Papers on Interdisciplinarity, Accountability and the Flow of Knowledge*. Oxon: Sean Kingston Publishing.

Strathern, Marilyn. 2016. "Inroads into Altruism". In *Releasing the Commons: Rethinking the Futures of the Commons*, edited by Ash Amin and Philip Howell, 161–177. London, New York: Routledge.

Talbot, Lyn, and Rae Walker. 2007. "Community Perspectives on the Impact of Policy Change on Linking Social Capital in a Rural Community". *Health & Place* 13 (2): 482–492. https://doi.org/10.1016/j.healthplace.2006.05.007.

Thomas, William I., and Florian Znaniecki. 1971. "A Polish Peasant Family". In *Peasants and Peasant Societies*, edited by Teodor Shanin, 23–29. London: Penguin Books.

Thompson, E. P. 1980. *The Making of the English Working Class*. London: Penguin.

Thompson, E. P. 1991. *Customs in Common*. London: Merlin Press.

Tooley, Beryl. 1985. *John Knowlittle: The Life of the Yarmouth Naturalist Arthur Henry Patterson*. Norwich: Wilson-Poole.

Tsing, Anna. 2015. *The Mushroom at the End of the World: On the Possibility of Life in Capitalist Ruins*. Princeton: Princeton University Press.

UK Government. 1988. *Norfolk and Suffolk Broads Act 1988*. www.legislation.gov.uk/ukpga/1988/4/contents.

UK Government, Department for Business, Innovation and Skills. 2012. "Industrial Strategy: UK Sector Analysis". BIS Economics Paper 18.

Vasudevan, Alex, Colin McFarlane, and Alex Jeffrey. 2008. "Spaces of Enclosure". *Geoforum* 39 (5): 1641–1646. https://doi.org/10.1016/j.geoforum.2008.03.001.

Walker, H. 2015. "Equality Without Equivalence: an anthropology of the common". London School of Economics. 6 May 2015. www.lse.ac.uk/newsandmedia/videoandaudio/channels/publiclecturesandevents/player.aspx?id=3104.

Weber, Max. 1978. *Economy and Society: An Outline of Interpretive Sociology*. Berkeley, CA : University of California Press, 1968. (2013 printing).

West, Patrick C., and Steven R. Brechin. 1991. *Resident Peoples and National Parks: Social Dilemmas and Strategies in International Conservation*. Tucson: University of Arizona Press.

White, Stephen D., and Richard T. Vann. 1983. "The Invention of English Individualism: Alan Macfarlane and the Modernization of Pre-Modern England". *Social History* 8 (3): 345–363.

Whittle, Jane. 2000. *The Development of Agrarian Capitalism*. Oxford: Oxford University Press. www.oxfordscholarship.com/view/10.1093/acprof:oso/9780198208426.001.0001/acprof-9780198208426.

Williamson, Tom. 1997. *The Norfolk Broads: A Landscape History*. 1st edition. Manchester, UK; New York, NY: Manchester University Press.

Williamson, Tom. 2000. "Understanding Enclosure". *Landscapes* 1 (1): 56–79. https://doi.org/10.1179/lan.2000.1.1.56.

Winkler, Justin. 2005. "The Eye and the Hand: Professional Sensitivity and the Idea of an Aesthetics of Work on the Land". *Contemporary Aesthetics* 3. www.contempaesthetics.org/newvolume/pages/ article.php?articleID¼289.

Winter, Michael, and Matt Lobley. 2016. "Is There a Future for the Small Family Farm in the UK?" Report to The Princes' Countryside Fund ISBN 978–902 746–36–7. London: Prince's Countryside Fund.

Wolf, Eric R. 1956. "Aspects of Group Relations in a Complex Society: Mexico". *American Anthropologist* 58 (6): 1065–1078.

Wolf, Eric R. 1966. *Peasants*. Englewood Cliffs, NJ: Prentice-Hall.

Wright, J. 2016. *A Natural History of the Hedgerow and Ditches, Dykes and Dry Stone Walls*. London: Profile Books.

4 Can you learn common sense?

One can fairly define common sense – as a vernacular concept in use in the English countryside – as a kind of pragmatic attitude, appropriate to a given shared working environment or "taskscape". The taskscape of the Broads National Park – continually reproduced by the environmental land management of farmers, conservationists, gamekeepers, and others – unfolds in accord with a specific, highly moral aesthetic of proximity. Due to the structuring power of this aesthetic for the farming community, it is fair to refer to this – as other social scientists have done – as a distinct "habitus of the common". But how do land managers acquire this habitus? Is it possible – despite popular expectations to the contrary – to learn common sense?

Overview: Strumpshaw Fen as a place of desire

RSPB Strumpshaw Fen lies at the heart of the Mid Yare Valley National Nature Reserve (NNR), a network of interlocking conservation areas lying in the flood plain on both sides of the River Yare, between the villages of Postwick and Reedham. The constituent reserves – Strumpshaw Fen, Buckenham Marshes, Cantley Marshes, Surlingham Church Marsh, Surlingham Broad, Brickyard Marsh, and Rockland Broad – are cared for by the Royal Society for the Protection of Birds (RSPB), one of the UK's largest and most influential conservation charities. Although the RSPB owns much of this land, a large part of Strumpshaw Fen – including the main visitors' centre, site office, workshop, and volunteer accommodation – is rented on a long-term lease from Strumpshaw Hall, a neighbouring estate. Strumpshaw Fen is the main point of arrival for visitors seeking to explore the valley; the Fen and its neighbouring fields and woodlands are criss-crossed by three main trails, which link together three hides, two pond-dipping platforms, and a nectar garden – all carefully signposted with waymarkers and interpretation boards, helping guide the visitors intellectually and physically through the Fen.

Key to this navigation process for the wardens was the notion of *succession*. *Succession* is the process by which open water – of the kind found in a broad or dyke – would gradually be colonised by, and altered by, sequential waves of plant growth. Each of the main habitats for which Strumpshaw Fen was known – reed swamp, valley fen, alder and willow carr, and wet oak woodland – represents different phases of succession. The open water at the edge of a broad would be first colonised by reeds (esp. *Phragmites australis*), which are highly tolerant of waterlogged soils. Each year, new growth from the rhizomic root systems formed by the reeds would displace the old, which would fall to the ground as litter. This litter would then partially decompose, forming a mat of peat that would in turn be colonised by other water-tolerant species – including sedges (genus *Cladium*), yellow loosestrife (*Lysimachia vulgaris*), rushes (genus *Typha*), and hemp agrimony (*Eupatorium cannabinum*). The drier soils produced by such fen communities would in turn provide a foothold for wet-tolerant trees such as goat willow (*Salix caprea*), crack willow (*Salix fragilis*), and alder (*Alnus glutinosa*). Such species would draw large amounts of water from the sub-surface, creating yet drier conditions, favoured by woodland species – like silver birch (*Betula pendula*), aspen (*Populus tremula*), oak (esp. *Quercus robur*), ash (*Fraxinus excelsior*), and beech (*Fagus sylvatica*).[1]

The goal of much of the management activity on the reserve was to interrupt this process from unfolding as it would – according to the wardens – under "natural" conditions. Without human intervention, the diverse range of habitats would in a matter of decades go through the full process of succession, and much of the reserve would revert to alder scrub or even oak woodland. O explained that, in prehistoric times, the periodic flooding of rivers like the Yare across the entirety of Britain would have created space in the primordial oak wildwood for patches of reedswamp, calcareous fen, and wet meadow to grow up. But with the rivers of Britain straightened, dredged, and tamed – held behind floodwalls and sluices, and held down by weirs – these marsh-making, erosive processes are now held in abeyance. As such, the artificial (that is, human) fabrication of those processes has become a necessity, if wetland plant communities are to survive. The wardens, contractors,[2] and volunteers who helped manage the reserve would spend each year coppicing invasive trees, cutting back and burning marsh and fen vegetation, digging out ponds, and dredging ditches. These labours stopped the reserve from becoming oak woodland, which was, in the words of O, "What it wants to be." T, one of the wardens on the reserve, observed on two occasions that the acres of reedbed and fen may have seemed like a fragment of untouched wilderness, but that they were "about as unnatural as you could possibly get," with the entire reserve being "a big garden … it's all managed" (See Strang 2009, 119–158 and Figure 4.1).

This need for careful management applies to fenland, but also to grazing marsh – a habitat that covered much of the northern bank of the Yare, from Strumpshaw meadows, through Buckenham Marsh, to Cantley in the far south of the reserve. Grazing marshes are a form of wet grassland where the water has been partly drained from the soil into ditches, creating pasture for herds of cattle. The pure water of the ditches makes them hotspots for

Figure 4.1 Reed cutting. Historical forms of land management that prevent succession were a major aspect of the RSPB's work in the Mid Yare Valley.

biodiversity, inhabited by many rare plants and invertebrates, like water soldier (genus *Stratiotes*) and the great diving beetle (*Dysticus marginalis*). The fields themselves are favoured nesting grounds for breeding waders – especially lapwings (*Vanellus vanellus*) – in the summer, and provide winter grazing for migrants like bean geese (*Anser fabalis*) in the winter.

In addition to maintaining the gates and ditches, a wide range of other tasks occupied H and A, the two wardens responsible for caring for the marshes. The *sward* – the community of grasses growing in each field – had to be carefully measured, and then grazed and mowed to specific lengths to provide ideal conditions for different species. Weeds – such as creeping thistles (*Cirsium arvense*) and spear thistles (*Cirsium vulgare*) – had to be removed, either by being dug up with a mattock, or through "topping" – the topical use of herbicide.[3] Fields that were not being grazed would usually be cut to produce a hay crop; harvesting this was a major undertaking. Breeding wader nests are highly vulnerable to predation, so during the breeding season (February–April), the number of crows (*Corvus corone*) and foxes (*Vulpes vulpes*) had to be monitored and controlled, with Larsen traps, the removal of nests, and hunting with guns by contractors. In both cases, the role of labour in "holding back" the land "from what it wants to do" was consistent.

This phraseology – of the land as "wanting to do" some things, which the wardens and workers worked hard to prevent – demonstrates sensitivity to the agency of the landscape among conservationists, and bears a similarity to the discourse of "good farming", discussed in Chapter 2 above. Just as farmers looked to neatly ploughed fields, clear of invasive vegetation as indicators of effective management inseparable from the social history of the place, the wardens and volunteers of Strumpshaw regarded tracts of scrub-free reedbed, a well-maintained path, or a marsh full of lapwings in the same way – all evidence of "good conservation".[4] As V, the Area Manager based at the reserve put it, "In the Broads context, we're farmers. We're just farming different things to other people. We're landowners...." Although conservationists would sometimes be at odds with farmers in the Broads, there was a tremendous degree to which practices and outlook were shared by both. Indeed, just as the RSPB were farmers in the Broads – after all, the grazing marshes were cut for hay or rented out for cattle grazing – many of Broadland's farmers were committed conservationists.

In Chapter 3 we have seen that the common working landscape of the Broads engenders a specific habitus, which in turn reinforces a particular set of common social norms that underscore the importance of an engaged, reasonable, pragmatic – "common sense" – attitude. Even those not directly engaged in land management retain the notion of the common as a "root metaphor" for social life, albeit one that is not habituated by their day-to-day practices. But the habitus of the commons – engendered through working and sharing the landscape – nonetheless clashes with a habitus of enclosure – engendered through the alienation or ownership of it. This chapter will examine both these kinds of habitus in greater depth, through my experiences at RSPB Strumpshaw Fen. As a site that is managed by a team of wardens and volunteers, as well as being a place frequented by recreational visitors, both kinds of Broadland habitus manifest together at Strumpshaw Fen, making it a perfect site to compare them. This comparison allows us to trace the limitations of the deep affection held by seasoned visitors – so-called Broadland Consciousness (O'Riordan 1969) – in reaching wider society, where the flood plains represent a "psychological barrier" rather than a place of common experience (Cocker 2008). The overall goal of this discussion will be to illustrate – in line with my broader effort to explore the symbolic force and meanings attached to "common sense" in English social life – the broader socio-ecological significance of sensory experience in English society.

Underview: thicket description of working your way through the landscape

If you were to find yourself beside the River Yare on a sunny day, watching the white boats and sailing yachts schoon across the water, you might be fooled into thinking that this shimmering waterway was the heart of the Mid Yare NNR. The gentle bevel of the flood plain is the Yare's handiwork, and the moods of this still-tidal river do have a significant impact upon the surrounding land. But the river is in fact sealed off from much of the fen and marsh on either side.[5] It is only through the medium of electric pumps and sluices – usually switched off or closed in the summer months – that water moves from the wetland into the river. Under most circumstances, the wetting of the plain is not the work of the river at all, but rather of fresh alkaline springs – low in nutrients and largely unpolluted, ideal for watering highly sensitive fenland ecosystems. The richest fens and channels, boasting the greatest diversity of plants and invertebrates, are therefore found tucked away under the skirts of the upland, where the springs are. The water of the Yare, meanwhile, is filled with effluent from fields and homes within its catchment, particularly the city of Norwich just upstream. Development has not just altered the chemistry of the river, but its very structure, too – the Yare's once meandering, shallow course has been straightened and deepened to better accommodate boat traffic, and to decrease the incidence of flooding. Today, the wardens at the NNR take pains to ensure that this polluted water doesn't flow from the river into the reserve. The only exceptions to this are times of drought, or flood, at which time the reserve acts as an overflow, containing water that would otherwise drown the settlements of Rockland Saint Mary, Brundall, and Claxton. When it behaves itself, therefore, the Yare is simply the recipient of any excess water in the drainage systems of the NNR. When it does not, it takes the role of a source of incipient risks – from flooding, to pollution events, to spreading non-native species. The river is also managed by outside agencies – namely, the Broads Authority and the Environment Agency – and not local landowners. Perhaps even more surprisingly, RSPB Strumpshaw Fen was closed to river traffic during the time of my fieldwork. This means that boaters move through, and not into, the reserve. As such, the Yare is oddly peripheral to the NNR that bears its name. It is a concourse, a pollution source, a risk vector, a boundary, a name, and a drain.

If you were to visit Strumpshaw Fen, therefore, the Yare would not be your way in. You would arrive by coming down Low Road, past open fields of wheat and salad crops, interspersed with lines of trees sheltering idyllic cottages. Upon entering the reserve, you would find the main visitors area, including the reception hide where tickets are purchased, a small pond and seating area, and the shady path that leads out onto the reserve. To the left of this is the workshop, which also houses toilets, a drying room, and a small office. Just beyond the workshop is the main office building, and a storage shed used by the visitor engagement team. Directly facing the crossing is the residential volunteers' cottage – where I lived for some four months. The visitor area is clearly demarcated from the main yard, onto which the cottage, the workshop, and the office all face, by a stout wooden fence. This demarcation of labour – between site management, and visitor engagement – was instructive, for it mirrors broader national discourses and debates about how land management and the general public fit together. The site management volunteers on the reserve repeatedly stressed how taking part in task days – essentially jumping the partition between the public side of the reserve experienced by visitors and the management side, allowed them to "see parts of the reserve you wouldn't get to see otherwise".

Geologically, the reserve is almost exclusively peat. The wardens enthusiastically impressed the depth of this soft, dark layer of waterlogged organic matter upon volunteers, by thrusting the handles of muck forks – which could be up to 1.5 metres in length – straight down into the ground with little effort, leaving little of the handle itself above the surface. Although the peat was largely safe to walk upon, in some parts of the reserve thick rafts of plant matter had grown out across what was once open water, creating a thick layer of sludgy liquid underneath the matted roots. Walking on "hover" of this kind was an unnerving experience. What appeared to be solid ground was often nothing of the sort. Leaping up and down would cause the earth to ripple and tremble out in every direction. This meant that, while walking on hover, you had to be sure to check the strength of the ground ahead before you put your weight upon it; holes were common, and if you trod in them without realising, you'd fall straight through into the watery sludge beneath, deeper than a man is tall. These holes would lurk unseen, even in plain sight. While out on the fen one day with a working party who were cutting reed, I saw one of the volunteers stumble into one, and go down up to his waist. Even the drier parts of the reserve were criss-crossed by old ditches, covered over with vegetation, and so they could surprise the inexperienced. It isn't hard to see how, in the past, people travelling across undrained land feared unquiet wetland spirits, who liked to lure humans off safe paths with spectral lights, before swallowing them up in the hungry mire (Porter 1969, 64–65; Hall and Coles 1994, 1).

This peat layer is the reason for, and is in turn caused by, the reserve's internationally significant ecology. Arising from it is a variety of different wetland habitats for which the Broads are known, including reedbed, fen, carr, wet hay meadows, and grazing marshes. Each of these habitats has a distinctive character. Reedbed – also known as reedswamp (for reedbed is a subtype of swamp) – is an area where the water table is at or above ground level for most of the year, allowing it to be dominated by common reed (*Phragmites australis*). Fen, by contrast, tends to be far more biodiverse than reedbed, with reeds and sedges being joined by a range of other plant species. Biodiversity in fenland ecosystems depends significantly upon water quality. Unlike either reedbed or fen, which are largely open to the skies, the thick, tangled wet woodland of a carr landscape is characterised by small trees such as alder, birch, and willow, and shrubs such as guelder rose (*Viburnum opulus*) and bramble (*Rubus fruticosus*). Grazing marsh and wet hay meadows are different again – these habitats are partially drained by ditches and more heavily grazed, and so they are dominated by wet-tolerant grassland species. With this diversity of habitats and the ease of access for visitors, especially at Strumpshaw Fen, the Mid Yare NNR is known as an excellent place for members of the public to see wildlife distinctive to the Broads – including marsh harriers (*Circus aeruginosus*), bearded reedlings (*Panurus biarmicus*), bitterns (*Botaurus stellaris*), Norfolk Hawker dragonflies (*Aeshna isoceles*), and Swallowtail butterflies (*Papilio machaon britannicus*).

But the wildlife of the fen, carr, and marsh were not altogether peaceable in their dwelling. Biting insects breed in the ditches and long grasses throughout the summer. The clouds of mosquitoes at dusk in the woods were so bad I couldn't venture out of the house after 6 pm in June. Wasps would frequently nest in the fen and could swarm and attack anyone who came too close. Saw sedge (*Gahnia aspera*) was so called for its serrated leaves that could cut through a leather glove if you grasped it. Crack willow (*Salix fragilis*) would drop its branches without warning. Highland cattle, much more aggressive than other breeds, would charge and could gore you if provoked. Even the reeds posed risks of their own: cutting your hands if you didn't wear gloves, and sheltering the fen from the wind, creating a stifling layer of humid air. Wet leaves presented a slip hazard, dead wood a trip hazard. One of the most conspicuous traits of the fen wildlife was a rapid rate of growth. Every Friday, I was tasked with walking around the reserve equipped with some shears, a lopper, and – occasionally – a petrol-powered brushcutter. These I'd use to cut back stray vegetation to keep the path clear. During the growing season, when the reedbed turned bright green with fresh leaves, it was literally all I could do to keep the brambles, nettles (*Urtica dioica*), reeds, and other shrubs from blocking the path completely. But this presented me with a clear view of how competitive all the plants were with one another. When the reeds had grown up, they were soon choked and weighed down by fantastic amounts of bindweed (*Convolvulus arvensis*) – a climbing plant – whose broad leaves and large white flowers weighed down the reeds so much that on gusty days great mats of vegetation would fall into the path, slowing my passage around the reserve as I had to stop to cut them up – an unending wave of greenery, breaking over weeks, fuelled by sunlight and water (see Figure 4.2).

The swarming of mosquitoes, the growth of saw sedge and reed, the cracking of willow,[6] the ardour of cattle and – especially – the wobbling, hole-studded hover and the blockage of paths by bindweed – all these risky, obstructive qualities highlight the fecundity and flux of the wetland, the vigorous, unpredictable interplay of violent life and gelid geology (McClean 2011). The unstable peaty ground and the exuberant, ever-changing biota combine to lend a unique quality to the experience of moving through the

Figure 4.2 Overgrowth. Keeping the paths and boardwalks clear of vegetation was a source of constant work for wardens during the spring and summer.

wetland areas in the Broads. Unseen holes, rapid regrowth, and herbaceous succession render movement not only difficult (Lund 2014), but also inherently unpredictable. An apparently sure footing may prove to be unstable when weight is applied; tracks that are clear one week are thick with briers the next. Maps are only ever a vague guide as to the presence of obstacles, and are no compensation for the hard-won experience of knowing the place through daily exposure to its varying caprices. The chaotic recalcitrance of geology and biology renders book-knowledge of little use when faced with the prospect of making your way through wetland.

It is in this fashion that I came to know the Mid Yare NNR. It resolved before me as a rich, rebellious community of plants and animals, all of whom had their own means of pursuing their various interests. As I worked on the reserve full time, this rough and tumble of their raw materiality scoured my own body – hefting chunks of wood knotted the muscles in my back; blades of sedge cut my hands and calloused my fingers; burning reed seared my arm hair and eyebrows; insect bites covered my skin as it hardened under the sun and rain; my blistered feet toughened up from the difficulty of seeking sure footings. All the while, the effort of making my way in the reserve whittled fat from my body and made me stronger. Much as Vergunst mused that his own boots are more evocative of his fieldwork than audio or video (Vergunst 2011, 210), the changes wrought on my own body were themselves a source of ethnographic insight (Jackson 1990; Stoller 1997). The tenor of my experience has, I suggest, taught me two primary lessons – the first being that practical activities have a profound connection with experience. The process of coming to understand a place is fundamentally changed when you've got to make a living there. Second, that the Yare Valley acted upon my body through *all* my senses. I was not merely observing the fen, or smelling it, or touching it – I was doing them all at once. The experiential roots of common sense, therefore, depend not just on one particular, privileged sense – such as vision, so pivotal in Western ocularcentrism (Feld 1997, 91–137) – but rather the entire sensorium (Howes 2003, XXII). If wardens merely watched, as one of my colleagues mused one day, they'd never get any work done. Echoing the Aristotelean concern for the integration of sensory experience (see Chapter 1), it is only when all the senses come together that the English common sense, as an attitude towards a common working environment, is possible.

But there is another factor to be considered – namely, the boundaries of the sensorium I attained, and thus the limit to the "common sense" it supports. I began this section describing the course of the River Yare, and how this riparian landscape is essentially sealed off from its river – a river that has now become largely a conduit for tourists, and waste water. While the sensorium attained via working on the land as I did certainly provided a unique and richly textured vantage-point on the Broadland environment, the sensoria acquired through touristic or recreational participation would necessarily be quite different. While tourists and other recreational users would consume the landscape temporarily *at a distance* – from a yacht deck, perhaps – in the spirit of *quiet enjoyment*, volunteers and wardens would perceive the landscape through an *aesthetics of proximity* (see Chapter 3), gained by *working their way through* the fen and marsh, coming face to face with the recalcitrant desires of the land itself; something that quiet enjoyment tends to ignore. The river bank is a clear barrier; that separates not just water, but also groups of people and patterns of sensory experience.

Counterview: quiet enjoyment and visitor experience

While historically the RSPB treated engaging the public as peripheral relative to conservation management and monitoring, today this situation is quite different. Visitor engagement was a critical part of the work done on the reserve, with half of the staff and most of the volunteers being occupied with assisting visitors. The RSPB's workers – both staff and volunteers – were divided into those concerned primarily with site management, and those whose time was directed towards welcoming members of the public (see Table 4.1). And, as with the Kabyle household discussed in Chapter 3, the layout of space at the site entrance – split between "the yard" that primarily supported the management of the reserve, and "reception" that supported visitors – translated the operational distinction between these two wings of the reserve's activities into the physical movement and bodily activities of the workers and visitors. This pattern of spatial organisation was repeated right across the landscape – both between the river and the reserve as described above, but also within the reserve itself. But even here, the very layout of the landscape was carefully managed to cater to the needs of visitors – even down to the growth of individual plants. Stands of willow in the fen were allowed to grow specifically to block the line of sight between each of the hides – all of which were in sight of one another – to create the impression of a wild, uninterrupted vista. As previously mentioned, plants were not allowed to grow in places where they would "get in the way" of the quiet enjoyment of visitors.

Employees are categorised according to their core responsibilities. The Thursday working party was a group of between three and six volunteers, most of whom were retired, who met every week and assisted the site management team. The Saturday working party met on the first Saturday of the month, and consisted of between ten and 20 volunteers, many of whom had full-time jobs or were in full-time education. Reception volunteers were also mostly retired, and took care of the front of house on the site, taking payment from the public, processing membership applications, updating the sightings board, and generally making visitors as welcome as possible. Monitoring volunteers were naturalists with expertise that the RSPB valued, and who undertook surveys of wildlife on the reserve as needed, especially during the

Table 4.1 The team at Strumpshaw Fen. A table showing the individual members of full-time staff (bold), part-time staff, and groups of volunteers (italics) based at Strumpshaw Fen

Site management	*Dual responsibilities*	*Visitor engagement*
H – Warden	**T – Warden**	Y – Volunteer Manager
O – Warden	**Z – Site Manager**	**R-l – Volunteer Coordinator**
A – Warden	**V – Area Manager**	H–h
Thursday working party	*Residential volunteers*	M
Saturday working party		J–
Monitoring volunteers		**D–n**
		S–e
		E–d
		Reception volunteers

breeding season. The broadest portfolio was held by the residential volunteers, usually young adults seeking experience of working in conservation, as part of their career development. They worked on a full-time basis, and were normally attached to the site management team, although in my time at Strumpshaw the visitor engagement team hosted a residential volunteer who supported the programme of events for families during the summer holidays. In return for their labour, residential volunteers received intensive training in a wide range of different conservation skills, accommodation on site for the duration of their placement, and assistance with finding future employment in the sector. Three of the wardens on the reserve had started out as residential volunteers.

Just as the workers on the reserve were classified into different groups, the visitors too were subdivided according to the kind of experience they sought. *Birders* were the core group of visitors to the reserve, easily noticeable by the equipment they'd carry – large pairs of binoculars and 'scopes – and the camouflage gear they'd wear. Keen birders who lived locally would often be regulars at the reserve, spending many hours sitting in the hides or at other viewpoints, equipped with sandwiches and thermos flasks of hot tea or coffee, watching the landscape quietly and patiently. Some birders would travel great distances to visit Strumpshaw at specific times of year to see rare species. News of "exotics" – birds that wouldn't normally be found on the reserve – would be posted on the reserve website, and would attract many birders from across the country.

Naturalists were interested in a different range of species to birders – particularly insects and plants. Naturalists would usually dispense with the big binoculars, and instead opt for smaller set, or a magnifying glass, paired with a notebook. Rather than viewing their specimens from a distance while standing still, they would slowly stalk around the paths of the reserve, studying the plants at the edges or the insects that lived off them. If a naturalist wanted to collect any specimens, she would need to get permission from the reserve. Birders and naturalists were viewed as the "core" and "traditional" members of the RSPB, and so were valued as being potential members of the charity. Both birders and naturalists would often build a long-term relationship with the wardens – all of whom were keen naturalists themselves – and, if they lived locally, would often assist the wardens in surveying the population of key species.

Photographers would look quite like birders and naturalists, except they'd be weighed down by a heavy bag full of camera equipment, and carrying a DSLR with a long zoom lens. Where the movement of birders was patient and smooth, those of photographers were more sudden and punctuated – rushing to take the perfect shot at the perfect moment.

Walkers usually didn't carry any equipment, apart from possibly a backpack or a walking stick. Unlike the categories of visitor so far described, walkers did not come to the reserve specially – they were either local people who treated the reserve as a pleasant place to stroll, or tourists who were visiting the Broads in general, rather than Strumpshaw Fen. One woman I spoke to had been coming there for years, but had not the remotest interest in the ecological significance of the place. She'd never even heard of a Swallowtail butterfly – the reserve's most notable species. Walkers would normally move through the reserve steadily, admiring the view from a distance as they went, usually not spending very long at any hide or viewpoint, preferring to rest at the benches along the route instead.

School groups would also occasionally visit the reserve, viewing the site according to the directions of a guide, taking down notes in the manner of naturalists.

Increasing strategic emphasis in recent years has been placed on the final group – *families*. Families were groups of visitors that included primary school-age children (4–11 years old), who would visit the reserve at weekends and during the holidays. They would normally stay closer to the entrance to the reserve, on a series of shorter circular paths with facilities dedicated to them. These included interpretation boards pitched at a younger audience, and activities like pond dipping, listening for bird-song, and making sculptures out of twigs. The engagement of families was a priority on the reserve, reflecting policy across the organisation. The RSPB argues that – unless people care about nature, and see the benefit of it – then there will simply be nobody with sufficient interest in birds or conservation more generally to support their work in the future (Royal Society for the Protection of Birds, Calouste Gulbenkian Foundation, and Green Exercise Research Team at the University of Essex 2013).[7] Emphasis was placed upon encouraging local children to care about wildlife. *Networks for Nature*, an initiative ongoing while I was on the reserve, involved RSPB staff and volunteers organising large public events and visiting schools within the NR13 postcode district to teach the children about the Broads and the importance of preserving its fragile ecosystem.

The important lesson to draw is that although all these groups have different kinds of experience on the reserve, these varieties of experience differ from that of actually working there. This is not to say that there was no overlap between visitors and workers – indeed, many of the volunteers were keen birders or naturalists who would visit the reserve when not volunteering, and wardens would often use the patterns of movement of birders and naturalists for survey work. But this overlap served to reveal the differences between the two sensoria, rather than indicate the lack of any divide. Volunteers themselves stressed the distinction, claiming that working "gave you a whole new perspective" on the place. Although the wardens and residential volunteers did occasionally use the visitor infrastructure as a base for doing surveys, survey work often required the wardens to head off the paths, and into the fen or across the marshes. Indeed, when we did use the hides to do one survey (of bittern nesting flights), one of the wardens said that it was "a bit of a luxury".

Each of the groups of visitors listed above shaped the management of the reserve in distinct ways that reflected their specific needs. The central concern of birders was quite straightforward – to get a good view of the birds on the reserve. The thick vegetation was cut back at regular intervals – both in front of hides, and at strategic points around the paths – to allow visitors to see into the fen itself. Hides allowed birders to sit and watch for extended periods, sheltering them from the weather while concealing them from the birds they wanted to see. Naturalists – including those with an interest in dragonflies, or rare plants – were catered to in a similar way: "meanders" were cut around pollen sources such as Buddleia (*Buddleja davidii*), viewpoints were cut onto particularly good habitats for certain invertebrates, and so on. Photographers could be more particular about the views on offer; some would complain about vegetation obscuring the shots they wished to take – despite the fact that the vegetation was what attracted the wildlife in the first place. Walkers preferred benches to hides, as their main aim was to get outside and enjoy the countryside in periods of clement weather. While birders and photographers merely sought out a good view, families were much more discerning – looking for excellent facilities and a great, all-round day out that their children would enjoy. The staff and volunteers on the reserve went to great efforts to cater to families' needs, organising events throughout the summer holidays to entertain children and their parents. While keen birders and naturalists were felt to be more inclined to "rough it" – to do without food, drink, and loos – it was felt by the RSPB staff that such a situation would not be acceptable to families. And all groups, it was felt, needed the paths around the reserve to be kept clear, free of troublesome local wildlife, and dry. As such, an important task was *maintaining visitor facilities* – a task that covered maintaining paths by keeping them clear of vegetation; ensuring any potholes were filled in; cleaning and repairing benches and hides; updating visitor information; putting up signs warning visitors of flooding or wasps' nests; cutting viewpoints; mowing and cutting back vegetation in the reserve's car parks; constructing pond dipping platforms; and repainting the inside of the Reception Hide. Although wardens often led such tasks, much more of their time overall would be spent on monitoring and maintaining the habitat.

Members of the public would often pass me by as I worked, looking very relaxed and at peace. When we talked to one another, they'd often remark at how lovely it must be to work outside in "the peace and quiet", and how "lucky I was" to work in a place like this. Such comments are illustrative, because they reveal the lens that many visitors to the reserve were applying – one of *quiet enjoyment*. From such a vantage point, even work in such a place becomes a form of leisure,[8] a perception strongly challenged by the embodied, harsh reality of actually doing the work. Indeed, these categories of visitors were contrasted with those who – in various ways – breached the official policy of the reserve, by engaging *too closely* with the site, in a way that caused damage or disturbance. Some of the regular birders I spoke to complained about one notorious case where a photographer trespassed into the centre of the fen to take photographs of a bittern nest. He got into difficulty, and had to be rescued by the wardens. "Unauthorised collection" – that is, harvesting animals and plants from the site – was strictly forbidden. In one instance, I was required to stop a woman from picking blackberries. Although there were clear pragmatic reasons for such restrictions,[9] an unintended effect of this policy was that the boundary between visitors and workers – between leisure and labour – was reaffirmed (see Figure 4.3).

So far, I have traced how visitor engagement, and visitors, the management of the site, and conservation are delineated. This division between site management and visitor engagement should not be taken as indicative of conservation being sacrificed for visitor engagement. Indeed, the evidence I collected would strongly indicate the contrary. The RSPB requires that all its staff complete timesheets to assess the amount of time they and the volunteers in their charge spend on all the varying tasks involved in running the reserve. This data is collated via a Content Management System (CMS). The Site Manager of Strumpshaw Fen, Z, made the data collected for the financial year from 2013–2014 (the year before I arrived) available to me. In 2013–2014, of the 14,901.25 hours of labour required by the NNR, 9,531.2 of those were supplied in exchange for wages ("staff hours"), while a further 3,971.3 were provided by volunteers ("volunteer hours") – either as part of residential volunteer placements, or by day volunteers. Almost 80 per cent of staff hours were taken up by conservation (including both monitoring and management tasks), with the rest being spent on either maintaining visitor

Figure 4.3 Photographers crowding around swallowtails. The gaze of the visitors at the reserve was highly concentrated upon specific vistas – such as a nectar garden full of butterflies (centre) – reinforcing the distinction between leisure and labour.

facilities or directly engaging the public. Volunteers spent slightly more on managing visitor facilities and public engagement (26 per cent). Overall, 77 per cent of all the work done by the staff and volunteers based at the NNR was directed towards habitat creation, monitoring, and administration, as opposed to public engagement. Clearly, most of the reserve's labour power, particularly its wage labour, was being directed towards conservation, rather than visitor engagement.

Compared to the relatively small proportion of the total labour power of the reserve it receives, visitor engagement has a large impact on the organisation of space on the reserve, and on the strategy of the RSPB as a whole (Clarke *et al.* 2013). The practice of maintaining visitor facilities was framed by RSPB staff as ensuring that visitors could access the site and experience the various animals and plants that lived there. But in rendering parts of the site accessible – paths, hides, and seating areas – wardens nonetheless reinforced the divide between the experience of the landscape they had, and the kind of experience enjoyed by the general population. While we were cutting back the brambles that encroached ferociously upon one of the fen paths, T joked that this sort of work meant that after a couple of months, all you could see as you walked along the paths was briers or leaves that needed cutting back. What to visitors would be a simple bank of verdure, or a lovely natural setting, was to workers on the reserve a living, material expression of present and future tasks – in short, a taskscape (Ingold 2000, 195). I was assured that – were the vegetation not cut back – within a month the path would be utterly impassable. This small admission highlights an important fact – all the work done by wardens to manage the reserve for visitors basically ensured that paying visitors *didn't need* to work their way through the reserve as volunteers and staff did. "What the land wants to do" as a result became less visible within the sensorium of visitors, held at bay by consistent maintenance by volunteers and wardens. Through this, visitors are free to experience *quiet enjoyment* of the reserve; watching, photographing, and walking about, but not through, the fens and marshes. Wardens, volunteers, and contractors, meanwhile, are constantly confronted with the wake, reality, and prospect of their labours, *working their way through* the landscape.

Interview: farmers, children, and the acquisition of common sense

The RSPB manages their estate in such a way as to build public engagement in the environment, and this enhances some kinds of sensory experience at the expense of others. The labours of the RSBP's staff and volunteers ends up reinforcing a distancing aesthetic of quiet enjoyment for visitors. This reveals a curious irony: in certain respects, the RSPB's efforts to heal the divide between people and the environment shores up the experiential aspects of that divide. Management itself, and the reason for it, ceases to be *sensible* for visitors. In the experiences of children within the *Networks for Nature* area

– the NR13 postcode district – we can see why the RSPB's efforts might end up having this effect.

Concurrently with my own fieldwork, Richard Irvine and Elsa Lee conducted a series of walks with the children of primary schools in the Broads area: from the villages of Horning, South Walsham, and Reedham (Irvine and Lee 2018; Irvine *et al.* forthcoming), part of a wider study of children's environmental attitudes across East Anglia (Irvine *et al.* 2016). All these villages are a short drive away from Strumpshaw – indeed, both Fairhaven School in South Walsham and Reedham School were part of *Networks for Nature*, and were visited by the RSPB over the same period as Irvine and Lee's research was taking place. The spontaneous comments made by the children from different communities about their surrounds as they walked through them are illuminating. The children took note of various features that highlight the role of land management in the local landscape (see Chapter 2) – recent thatching "done from local reed" (Irvine *et al.* forthcoming); local farming practices; the installation of flood defences; and the harvesting of blackberries with one's family (Irvine *et al.* forthcoming) – responses of this kind being particularly frequent in Reedham (Irvine *et al.* forthcoming), the most agricultural of the three villages. Recreational pursuits like fishing, "manhunt",[10] and boating – particularly in the village of Horning, which has a bustling river frontage – were also provided routes toward engagement with the landscape for the children. In these villages, the children voiced considerable anxiety regarding the spread of housing development over greenfield sites – reflecting a prominent anxiety among the adult population in the region, which was a frequent topic of conversation on the volunteer days at Strumpshaw.

Children in all the villages show a keen awareness of their landscape: being interested in, and clearly spending much of their time exploring and playing in what public spaces were open to them. But such engagements ceased at the edge of the wetland areas. This signifies that:

> the wetland has become a specific domain, defined by a particular division of labour which the children are not part of.... Just as the villages themselves hug the margins of the Broads Executive Area, but are not (technically) "in" the Broads, so too the children's experiences are of dwelling on the margins of wetland habitats. Fen and marsh are, in this sense, sealed off from the children just as they are from the river that appears to run through them. The core characteristics of the Broads' ecology are peripheral to the children's embodied experience of the landscape.
>
> (Irvine *et al.* forthcoming)

Irvine *et al.* argue, following Rautio (Rautio 2014, 462), that a "mingling" between children and the material world allows a shift from knowing *about* the world, to knowing *with* the world. They move to contest that "enclosure disrupts this mingling precisely by restricting movement, restricting every banal material encounters, and therefore placing children in a position where they are restricted in what they can 'know with'" (Irvine and Lee 2016, 4). I suggest that these "banal, material encounters" are central to how wardens and volunteers come to know the land they manage – but that visitors, including local children – by contrast and by merit of the "facility" of their access, have more constrained opportunities for such encounters.

An important example of how this kind of situated learning conditions the sensorium of wardens is how one learns to navigate the fen, when away from the paths accessible to visitors. On my first day on the reserve, O took me out to check on the three highland cows at Surlingham Church Marsh. We walked through the lush fen, the soft aromas of water mint and wet earth rising up from every footfall. To me, we were completely lost in a sea of scented green. O, however, had "got his eye in" (see Chapter 2). On grazed fen, he could see the ditches from a distance, marked by long lines of reeds – the plants having grown tall because the cows couldn't reach them. He knew exactly where he was. But even when equipped with this knowledge, I found it hard to see these lines. On subsequent trips to the fen looking for Himalayan balsam (*Impatiens glandulifera*), I couldn't "get my eye in" and see the ditches in the reeds, and got hopelessly lost. Everyone on the warden team was adamant about the importance of this level of familiarity with the landscape, spontaneously mentioning it during interviews and in ordinary conversation. When discussing an encounter with a visitor who was trespassing off the path at Strumpshaw, T explained that:

> [He] was standing about off a trail, about a foot away from a blackcap nest, with a blackcap absolutely shouting crazily at him.... And, and he obviously didn't have a clue, he was thinking "these lovely birds, singing away...." So he, he didn't have the natural, the naturalist's knowledge, that (i) he should stay on the path because birds nest on the floor, and (ii) you know, [laughs] he was right next to this blackcap that was absolutely berating him, and he was like ... "that's nice ...".

For H, familiarity with the land in which he worked – grazing marsh – arose less from "naturalists' knowledge", and more from the experience of working with cattle and gazing land. When asked about how he translated the management plan into action, H said:

> If you're looking at creating a particular type of sward it's to do with ... livestock management, so managing our stockman to put the right number of stock in the right place at the right time, or take them away, if they need to come away ... erm, which comes from experience really ... that's farming background ... grazing ... hard work ... erm, and there's like the other agricultural element to that is machinery use, and using toppers, mowers to manage that sward.

For H, book-learning was less important than gaining practical experience of pastoralism: watching "what a cow does", asking questions of those with whom you were working – such as graziers, stockmen, or the landowner. Though T and H's perspectives may strike a contrasting note – between "knowledge" and "experience", the commonality of their approach becomes clear when they are asked to define common sense. From T's interview:

> Doing what is required *without thinking or being taught* … but you can probably get better at. So someone with common sense, will almost, will do all the things you would expect them to do, without … y'know it's quite tricky, that one, it's a good'un! … it's basically *not doing anything stupid!* [laughs] Or, or not doing anything wrong unless, they believe it's the right thing to do for another reason….
>
> (Emphasis mine)

And from H's:

> JONATHAN: "Could you possibly define common sense for me?"
> H: [laughs] "Am I allowed to think about that one?"
> JONATHAN: "You can think as long as you like!"
> H: "I'm not sure common sense is *common* sense."
> JONATHAN: "No?"
> H: "It's not common to everybody – it depends on what you're doing – erm – common sense relies on your *experience and worldliness*…. And, if my experience and worldliness in one particular area is not the same as yours, then we probably come to…." [trails off]
> JONATHAN: "Different common senses?"
> H: "Different common senses! Yeah." [laughs]
>
> (Emphasis mine)

Here we get a sense of how "banal material encounters" inform the sensorium of wardens – a sensorium to which visitors have only partial access. When viewed together – especially with H's comments – these encounters reveal the assumption of a link between the development of competency in a particular landscape, and the honing of common sense. It is important to note that while T has a background of being a keen naturalist – something that brought him into conservation and still dominates of a lot of his spare time – H had more of a focus on the agricultural aspect of the reserve. Their professional and personal histories are reflected in their statements about common sense. T's view has a more intellectualist bent (concerned with knowledge and stupidity), while H's is more invested in professional experience. But T nonetheless tells us that common sense requires no thought, nor formal teaching, and H insists on its plurality, arising from a specificity tied to particular places. Using these insights to reflect upon my own failure to "get my eye in" to Surlingham Church Marsh, knowledge of how to navigate through the fen in theory was not sufficient, until it was inculcated at a deeper – attitudinal – level. This suggests that, although knowledge and experience are both important, as T and H attest, common sense – appropriate to the specific place where you are getting experience of working – is also of pivotal importance, and is even positioned as a proper foundation of both knowledge and experience. There's also a clear subtext here of normativity – as T's story about the blackcap makes quite explicit. Not having the sense common to a particular landscape leads you to behave in the wrong way.

At this juncture, it's important to return to the discussions of previous chapters. Farmers, as we saw, develop what Winkler calls

> an aesthetic of proximity [and] of the hand [that] does not necessarily have to leave out toil and suffering to create joyful experience, as is the case with the dominant panoramic aesthetics. *Enjoyment even does not emerge without toil and time spent on the task.*
>
> (Winkler 2005, emphasis mine)

As in the previous section, I would qualify Winkler's point to avoid romanticising work unduly, adding that "enjoyment" for those who work the land is often more akin to a kind of grim satisfaction at tasks completed, rather than "enjoyment" per se. This qualification notwithstanding, the dichotomy struck here is helpful; what Winkler calls a "panoramic aesthetic" is the mode of landscape engagement that lies behind the practise of *quiet enjoyment* by the visitors to the reserve. Wardens and volunteers, meanwhile, spend most of their time developing the opposite kind of "aesthetic" – one rooted in practical action, painful toil, and a social history of successive management decisions by known and named people. This demonstrates how a particular attitude to the world – the sense of the common – emerges for Broadland's land managers, ingrained through a collective habitus that structures (and is structured by) their working landscape. Visitors, by contrast, have a quite different habitus, that is both supported by and effaces the labour that goes in to the very places where it emerges.

In locating two, markedly distinct ways of relating to the landscape – the *quiet enjoyment* of visitors, and the *working through* of wardens – my study of Strumpshaw Fen has many similarities to Strang's case study of Kowanyama Aboriginal and White rancher perspectives on the Mitchell River catchment in Far North Queensland (Strang 1997). Strang argues that:

> human environmental interactions are largely an expression of cultural values; that these recur consistently through a range of interconnected cultural forms that, acting upon each other, maintain a coherent pattern of value; and that in articulation with a range of universal human imperatives and ecological pressures, this pattern of value creates a particular "mode" of interaction with the environment.
>
> (1997, 4)

In the context of the Mitchell River, Strang points out, Kowanyama and White communities have a radically different perception of their surroundings. The indigenous population see their country as densely textured, meaningful, and social, concretising relationships between humans and non-humans extending over generations, revealed through stories, memories, and place-names (Strang 1997, 105, 220; Ingold 2000, 116–121). This reflects the observation that, in indigenous Australian contexts more generally,

> all matter (human and animal bodies, objects, and environments) is conceived as the congealed labor of ancestral Dreaming beings. While the mythic actions of some dreamtime ancestors were concentrated at certain now-sacred sites, the land is more generally permeated by signs of their present-day intentionality and agency.
>
> (Povinelli 1995, 509)

In contrast to this holistic understanding of the landscape held by indigenous people on the continent, White Australian society, "in both cosmological and material terms, [is] characterised by immense fragmentation and abstraction" (Strang 1997, 283), which, Strang draws on Miller to suggest, are two components that together create the prevailing sense of alienation in modern societies (Miller 1994, 78). In conclusion, Strang advances a set of cultural attitudes that either encourage or discourage affective environmental values (see Table 4.2).

This list raises the possibility of societies and groups exhibiting varying degrees of affective attachment to place, based on the presence or absence of these factors. What I would argue here is that "common sense" – as experienced by farmers, conservationists, and other land managers in England – denotes several of the factors identified by Strang as encouraging affective connections with place; specifically activities relating directly to the local environment, moral structures related to the local environment, detailed knowledge of the immediate area, and cathection with one manageable area. These cultural values are themselves programmatic for certain practices and arrangements of space, and are in turn reinforced by those same practices. Many of the features Strang identifies strongly resemble the values of Broadland organicism (see Chapter 3), an outcome that is unsurprising if we consider the strong affective connection that many of those authors express for the Broads.

Teleview: "Broadland Consciousness" versus "Barrier Consciousness"

For those seeking to explore the Yare Valley in its present state of enclosure, the process is anything but straightforward. The nature writer Mark Cocker, in an intimate portrait of the Yare Valley, states that:

Table 4.2 Encouraging and discouraging factors in affective environmental values – selected from a more extensive list in the original

Encouraging factors	*Discouraging factors*
Inalienable land ownership	Alienable land ownership
Land considered to be unique	Land commoditised
Continuity of residence	Discontinuity of residence
Inhabitance of one specific area	Changes in location
Cognitive encompassment of one area	Attempts to "encompass" many areas
Cathection with one manageable area	Attempts to cathect with many/larger areas
Detailed knowledge of immediate area	Lack of knowledge of local environment
Long-term historical association with place	Lack of historical association with place
Socio-spatial forms mediated by place	Independent socio-spatial forms
Economic mode utilising local resources	Economic mode based on imposed resources
Economy aimed at sustainability	Economy aimed at growth/expansion
Material culture locally produced	Material culture imported
Activities relating directly to local environment	Activities relating to imposed elements
Geographically specific representations	Generic representations of landscape
Education focused on local environment	Education generic and wide-ranging
Intergenerational transmission of knowledge	Constant change separating generations

Source: Strang 1997, 287–288.

> I quickly found that [the Yare] held sway over an awkward, inaccessible landscape. In the entire stretch from Norwich to Yarmouth it's bridged at either end in the towns themselves, but in between there is no physical structure across it. I routinely meet people who live on one bank and who have never visited the village opposite, although they may look upon the place every day of their lives. Once I even encountered a couple during a walk at Buckenham who'd lived all their lives in a village on the north side of the river, and didn't even know the name of the village on the other side. *The Yare has become a psychological barrier.*
>
> (Cocker 2008, 20, emphasis mine)

This construction of the Yare as a "psychological barrier" stands in apparent contrast to the close and faithful devotion to the river and its environs expressed by the recreational visitors to Broadland, who number some eight million per year (the Broads Authority 2016). Among the occasional visitors, there are those who have a deep and abiding affection for the region, and then return year after year. These people possess what O'Riordan calls "Broadland Consciousness" (O'Riordan 1969, 44–45). As O'Riordan points out, this appears to be the product of a "sifting out" process, in which only "seasoned" Broads-lovers return to the Broads on a regular basis.

One example of this sort of consciousness is the introduction written by the actress Olivia Colman to the 2014 edition of *Broadcaster*, an annual visitors' guide to the Broads, published by the Authority. Colman recounts her idyllic childhood living in a cottage on the edge of Ranworth Broad, and at Horstead – an experience that was "lovely, feral, free". Colman explains that it's

> almost impossible to offer children that kind of experience today. But in the Broads, you still can! [She continues:] Wherever you are you're always aware of water – rivers, broads, smaller waterways and the coast. If you want to try sailing or canoeing it's an ideal place to start.... When I think of the Broads, I think of a great place to escape to, whether you want activities to keep children (and adults) entertained or you just want to sit on a tranquil riverbank taking in the stunning scenery.
>
> (Colman 2014, 3)

Colman's recreational perspective on the Broads does not notice anything about its character as a working, industrial landscape. Instead, going to the Broads is an opportunity for an "escape". Despite the fact that the scenery she admires is – as my ethnographic description in this chapter attests – entirely man-made, that making is not something she feels drawn to remark upon. This pattern is even more strongly expressed in *The Norfolk Broads* (Pullinger and Crawley 2014) – a promotional publication sponsored by Richardson's Boating Holidays, a private tour operator. While the numerous virtues and exquisite beauty of the landscape are extolled on page after page, marshmen are billed as a "dying breed" (ibid., 11), while wherries are euhemerised as "beloved survivors" (ibid., 12–13). The management of the Broads is squarely located in a traditional past, or left out entirely from recreational representations.

Despite the doubtless passion of those seasoned visitors who possess Broadland Consciousness, the very fact that so many people are "sifted out" reveals the fact that such consciousness has real limits. I would take the "sifting" further, and stress the extent to which those who don't belong to particular interest groups; those who aren't enthusiasts with a stake in this particular landscape have no reason to go there at all. This means that, for all those people who possess Broadland Consciousness in the region, there will be many more who do not. Furthermore, the way in which these visitor groups engage with the landscape is quite different from the close, sustained, practically oriented attention shown by the land managers I worked with. Instead of being a conduit for ordinary "banal experiences", as for those who work and worked there, for the enthusiast seeking quiet enjoyment, each broad is a destination – a beautiful place for fun and relaxation, set quite apart from daily life.

In the early 2000s, following the election of a Labour government in 1997 under the reformist Tony Blair, public spending was increased on average by 4.8 per cent in real terms annually (Fielding 2002; Chote *et al.* 2010, 4–5).

During this period of relatively abundant state funding, the Broads Authority could pursue initiatives that, it was hoped, would address the divide between the people and the landscape, and build trust in the Authority itself. As M, a former countryside warden explained:

> [Our role] was being a linchpoint [*sic*] in the community to pass messages across, pass messages back and forth, and also practically to manage the footpaths, to make sure access was open to welcome people out into the Broads. And also to educate people about the Broads. So it was a mixture of hands on, practical … and also, getting people out there, and making them understand, and working with communities.

However, M was clear that there was no overarching sense of people living in the Yare Valley communities having a sense of being connected to one another. Indeed, she reinforced Mark Cocker's commentary about the "psychological divide" represented by the river:

> I think that side of the river [gestures to the south side] has no connection to this side of the river – at all. But that is very recent, really, in recent memory because there used to be so many little foot ferries across – there used to probably be about, between here and Norwich, there probably used to be about six. Erm, and there was always talk about getting some of those back and running, and that kind of never happened, really. But in the past, I used to know a lovely lady, her dad owned a pub in Surlingham, and there used to be a ferry across to Brundall, and that used to be a regular journey. And that's not that long ago, she's not that old, she's in her 60s now. So, not within my time, but within living memory, this would have been a much more cohesive river valley.

This "Barrier Consciousness" didn't just hold people back from visiting the far side: in many cases, people steered clear of the river valleys entirely. In my time in East Norfolk, I was shocked by the number of long-term local residents who had either neither visited the Broads, or considered them as distant to their ordinary lives. One dairy farmer who lived in Broadland District had never been to the Broads, except one summer as a child when her family had rented a boat – an experience she only dimly recollected. Some people I met in Norwich – despite the Yare flowing through their city – didn't even know the National Park existed. One local resident in Brundall summarised this point very clearly, when he remarked:

> Do I think people are aware and value that they're in a national park? Locally here? I think they do if they sail, or if they're walkers. They don't have that much contact with it, outside of that, … They're not in it, because they're out on the edge of it.

As a recreational landscape, the rivers, lakes, and wetlands of the Broads can never be anything other than of partial interest for many people.

Conclusion: common sense and English sensoria

My aim here has been to strengthen the assertion made in the previous chapter that the common – a configuration of the landscape as an open, worked space – engenders a specific sort of habitus that contrasts sharply with the habitus entailed by inhabiting a landscape under a state of enclosure. This former configuration is still experienced in limited ways by land managers – both farmers and conservationists – in Broadland, and still has a powerful hold on the wider social imagination of England as a root metaphor. Enclosure and commoning are not binary states imposed in the landscape in an absolute way, but rather processes that can be experienced by individuals to varying degrees. Strumpshaw Fen is private property, and therefore has in a legal and historical sense already been long enclosed. However, my point here has been to indicate that even heavily privatised landscapes contain the potential for commoning. Wardens, like farmers, can and do treat their workplace as a common, to which they have certain shared rights and obligations, and where a "common sense" is possible. And yet, visitors, by contrast, have far fewer opportunities for "commoning" of this kind on the reserve, and see the land quite differently as a result. This variety in relative levels of enclosure and commoning is revealed by different modes of aesthetic and physical engagement with the place – with RSPB staff and volunteers holding an aesthetic based on work and proximity, while the visitors opt more towards quiet enjoyment. Visitors can secure views into the reserve from the paths or the river, while those who work there see through it all.

The conventional mode of viewing the landscape as mere scenery is critiqued by Strang, who suggests that for White people in her fieldsite, "the land is a stage for human activities, rather than a medium of organisation, and the available roles could be performed interchangeably in any similar economic or social situation" (Strang 1997, 280). My suggestion here is that this critique is itself something that can be situated within broader debates and anxieties in English-speaking cultures: a desire to situate a common sense engagement with the physicality of the landscape against the distancing aesthetic of quiet enjoyment, with the former being a solution to the ecological problems signified by, if not directly caused by, the latter. This intervention finds eloquent expression in recent climate change discourse:

> Because this is a crisis that is, by its nature, slow moving and intensely place based. In the early stages, and in between the wrenching disasters, climate is about an early blooming of a particular flower, an unusually thin layer of ice on a lake, the late arrival of a migratory bird – noticing these small changes requires the kind of communion that comes from *knowing a place deeply, not just as scenery but also as sustenance, and when local*

> *knowledge is passed on with a sense of sacred trust from one generation to the next.* How many of us still live like that? Similarly, climate change is also about the inescapable impact of the actions of past generations not just on the present, but on generations in the future. These time frames are a language that has become foreign to a great many of us. Indeed, Western Culture has worked very hard to erase indigenous cosmologies that call on the past and the future to interrogate present-day actions, with long dead ancestors always present, alongside the generations yet to come. In short, more bad timing.
>
> (Klein 2014, 159, emphasis mine)

Within Klein's polemic rests an important observation, which is concordant with the material presented in the previous two chapters – both from the spontaneous utterances of school children collected by Irvine and Lee, and from my own experiences and those of my colleagues in the Mid Yare NNR. Klein identifies "knowing a place deeply" as knowing it as *both* scenery and sustenance, an exhortation of the importance of seeing beyond the beauty of nature, through a pragmatic engagement with it; from pleasant views, to banal material encounters. It is precisely this sort of synthesis that the "Barrier Consciousness" described above precludes. Klein's words chime very strongly with the sentiment of the wardens regarding common sense; although they did not subsist off the landscape of the fen, they were undoubtedly engaged in the practical, difficult business of growing a "crop" from it. Thinking of engagement with the landscape in terms of component practices and values that can be independently present or absent from one another, but that work in concert to create different habits and notions of sensibleness, helps us to better understand the complex relationship between scenery and sustenance. Rather than thinking of the struggle between enclosure and commoning of the landscape as an either/or, the sensoria of English society today demonstrate the extent to which these twin processes are both gradual in nature.

Notes

1 For full details of the botany of British fenland, see Mabey 1998. For a detailed survey of the botany of the Broads specifically, see George 1992.
2 The reserve employed contractors to undertake certain major works, or to support the warden team.
3 Herbicide (namely glyphosate) was very strictly controlled and administered with great care on the reserve.
4 I observed my conservationist friends studying and commenting critically (or positively) on the management of the countryside we travelled through in much the same way as farmers would: noting hedges cut at the wrong time of year, overgrown footpaths, or bare fields as signs of "bad conservation" by the landowner.
5 It is important to note that these seals are not perfect, however – a limited amount of seepage from the river into the reserve was unavoidable.
6 The crack willow's dropping of twigs and branches is actually a way of reproducing. Because crack willows grow near water, any twigs that fall in this medium

will be carried downstream to distant locations where they may take root if they are washed ashore.

7 The observations and arguments made in this report are based on research completed by the University of Exeter and the Calouste Gulbenkian Foundation.

8 Some visitors I spoke to, erroneously believed that wardens were "paid to watch birds".

9 Collecting birds' eggs and picking wild flowers remain criminal offences in the UK, and the RSPB has been assiduous in bringing prosecutions for these and other wildlife crimes (Bibby 2003; the Farmers Guardian 2004). Picking wild berries and harvesting wild food plants – though a legal practice that the wardens viewed as laudable – was not permitted on the reserve because it might deprive wild animals of food sources.

10 A two-team children's game that is a variant of hide-and-seek played in the dark – where one group of players (or sometimes, an individual player, who is "It") seeks to find members of the other group.

References

Bibby, Colin J. 2003. "Fifty Years of Bird Study". *Bird Study* 50 (3): 194–210. https://doi.org/10.1080/00063650309461314.

Broads Authority. 2016. "Facts and Figures". National Park Authority. The Broads Authority. 2016. www.broads-authority.gov.uk/learning/facts-and-figures.

Chote, Robert, Rowena Crawford, Carl Emmerson, and Gemma Tetlow. 2010. "Public Spending Under Labour". 2010 Election Briefing Note 5. London: Institute for Fiscal Studies.

Clarke, Mike, Steve Ormerod, and Danae Sheehan. 2013. "RSPB Annual Review 2012–2013". Annual Report. United Kingdom: Royal Society for the Protection of Birds.

Cocker, Mark. 2008. *Crow Country*. London: Vintage.

Colman, Olivia. 2014. "Welcome to the Broads". *Broadcaster*, 2014.

Farmers Guardian. 2004. "UK Wildlife Crime Data". *Farmers Guardian*, 4.

Feld, Steven. 1997. "Waterfalls of Song: An Acoustemology of Place Resounding in Bosavi, Papua New Guinea". In *Senses of Place*, edited by Steven Feld and Keith Basso. Santa Fe: School of American Research Press.

Fielding, Steven. 2002. "'New' Labour and the 'New' Labour History". *Moving the Social* 27: 35–50.

George, Martin. 1992. *Land Use, Ecology and Conservation of Broadland*. Chichester: Packard.

Hall, D., and J. Coles. 1994. *Fenland Survey: An Essay in Landscape and Persistence*. UK: English Heritage.

Howes, David. 2003. *Sensual Relations: Engaging the Senses in Culture and Social Theory/ David Howes*. Ann Arbor, MI: University of Michigan Press.

Ingold, Tim. 2000. *The Perception of the Environment: Essays on Livelihood, Dwelling and Skill*. London: Psychology Press.

Irvine, Richard, and Elsa Lee. 2018. "Over and Under: Children Navigating Terrain in the East Anglian Fenlands". *Children's Geographies* 16 (4): 380–392. https://doi.org/10.1080/14733285.2017.1344768.

Irvine, Richard, Elsa Lee, Laura Misseldene, and Jonathan Woolley. Forthcoming. "Moving and Dwelling: Children at Home in the Norfolk Broads". *Sociologia Ruralis*.

Irvine, Richard, Elsa Lee, Miranda Strubel, and Barbara Bodenhorn. 2016. "Exclusion and Reappropriation: Experiences of Contemporary Enclosure among Children in Three East Anglian Schools". *Environment and Planning D* 34 (5): 935–953. https://doi.org/10.17863/CAM.22.

Jackson, J. 1990. " 'I Am a Fieldnote' Fieldnotes as a Symbol of Professional Identity". In *Fieldnotes: The Makings of Anthropology*, edited by R. Sanjek. Ithaca and London: Cornell University Press.

Klein, N. 2014. *This Changes Everything: Capitalism vs. the Climate*. New York, NY: Simon and Schuster.

Lund, Katrin. 2014. "Landscapes and Narratives: Compositions and the Walking Body". *Landscape Research* 37 (2): 225–237.

Mabey, R. 1998. *Flora Britannica*. London: Chatto & Windus.

McClean, S. 2011. "BLACK GOO: Forceful Encounters with Matter in Europe's Muddy Margins". *Cultural Anthropology* 26 (4): 589–619.

Miller, Daniel. 1994. *Material Culture and Mass Consumption*. Oxford: Blackwell.

O'Riordan, Timothy. 1969. "Planning to Improve Environmental Capacity: A Case Study in Broadland". *The Town Planning Review* 40 (1): 39–58.

Porter, E. 1969. *Cambridgeshire Customs and Folklore*. London: Routledge and Kegan Paul.

Povinelli, E. 1995. "Do Rocks Listen? The Cultural Politics of Apprehending Australian Aboriginal Law". *American Anthropologist* 97 (3): 505–518.

Pullinger, Stephen, and Bob Crawley, eds. 2014. *The Norfolk Broads – Our Heritage, Our Future – a Unique Perspective*. Norwich: Eastern Daily Press.

Rautio, Pauliina. 2014. "Mingling and Imitating in Producing Spaces for Knowing and Being: Insights from a Finnish Study of Child–Matter Intra-Action". *Childhood* 21 (4): 461–474. https://doi.org/10.1177/0907568213496653.

Royal Society for the Protection of Birds, Calouste Gulbenkian Foundation, and Green Exercise Research Team at the University of Essex. 2013. "Connecting with Nature – Finding out How Connected to Nature the UK's Children Are". Sandy, Bedfordshire: Royal Society for the Protection of Birds.

Stoller, P. 1997. *Sensuous Scholarship*. Philadelphia: University of Pennsylvania Press.

Strang, Veronica. 1997. *Uncommon Ground: Cultural Landscapes and Environmental Values*. Oxford: Berg.

Strang, Veronica. 2009. *Gardening the World: Agency, Identity and the Ownership of Water*. New York, NY: Berghahn Books.

Vergunst, J. 2011. "Technology and Technique in a Useful Ethnography of Movement". *Mobilities* 6 (2): 203–219.

Winkler, Justin. 2005. "The Eye and the Hand: Professional Sensitivity and the Idea of an Aesthetics of Work on the Land". *Contemporary Aesthetics* 3. www.contempaesthetics.org/newvolume/pages/ article.php?articleID¼289.

5 Why is common sense so scarce?

The winter is a time for cutting things down. Most days from October to March you'd see a long column of smoke rising into the wide Norfolk sky above Strumpshaw Fen, as dry wetland vegetation was piled up and burned. On the first Saturday in February 2016, I returned to Strumpshaw to go out with the Saturday Working Party. That day, we were tasked with cutting back willow scrub. Before too long, Bill and I were standing under a grim bank of alder at the edge of the reserve, stacking up long spindly wisps of willow that the rest of the group had been cutting down, out on the fen. It was steady, satisfying work, which became steadily more satisfying as we got the hang of it, experimenting and trying different ways of getting the willow branches to lie together. Initially, we just threw the branches as far as we could between the alders, into the mud, but over time our technique became more refined. The circumstances were constantly changing: the speed with which the branches were brought to us, the amount of space available, our own levels of energy, the size of the branches. We figured out that weaving the branches into those that were already stacked was the most efficient way to go about it; the thinner twigs broke off as you turned and snaked each branch in, meaning that each piece of brash took up less space, and required less energy to dispose of than trying to throw them further and further into the wood. We kept an eye on each other, and paid loose attention to the wood and the work, while chatting about other things.

Bill asked me about my thesis. I had decided to write it about common sense, I told him. He laughed. "Oh dear!" He exclaimed. "That must be difficult; there's not a lot of that around, is there?"

* * *

The joke that common sense is scarce was a common reaction to my chosen thesis topic. Indeed, I began to consciously crack this joke myself to English people I met, in Norfolk and back in Cambridge, to gauge their reaction. The result was always a laugh, or at least a smile. The ironic scarcity of (supposedly) "common" sense was clearly a familiar refrain to English people. Humour can offer an invaluable insight into norms and paradoxes (Swinkels and de Koning 2016, 8) – in this case, the ample eye-rolling and knowing looks were indicative of what was so funny: this observation was funny because it was true. English people believe that common sense *actually is* rather scarce. Not only is this ironic, but it is a sore point – a graze on the English body politic, wrapped in bandages of humour. Common sense – although it *should* be "common" – is lamentably absent from society.

In Chapter 3, I argued that the common serves as a crucial root metaphor in English culture, that it summarises prevailing views about the organic, pragmatic, and sensible nature of social norms. In Chapter 4, I explored the role of "sensibleness" – wherein a common landscape acts as a medium into which those norms are materialised through a particular habitus for those who manage it. The claim that common sense is rather rare articulates these attitudes in two ways. First, when used in the form of a joke (as above), we see how it is inclusive; it invites the speaker and the listener to see themselves as part of the same moral community of "sensible people", opposed to those who aren't. When it's used in the second sense – as a criticism – the emphasis is upon exclusion, picking out specific persons or classes of persons who lack this basic quality. These two uses cut in contradictory directions: the joke pits specific sensible individuals against a prevailing (and ironic) lack of common sense in society at large, while the criticism identifies particular agents or actions that reveal a lack of a sense common to a general, unspecified, "sensible" audience, pointing out errors of judgement that would be obvious to any right-thinking person. My goal here is not to resolve this contradiction – common sense means different, even contradictory, things to different people on different occasions – but rather to use this tension to open up an important feature of Broadland society, namely, controversies over the management of its landscape. Within these controversies, the absence of common sense – in both its joking and critical guises – plays a key role. This absence, as I will go on to suggest, reveals further dimensions to the paradoxical relationship between the common, and the state of possessive individualism – and its "habitus of enclosure" – that governs the English landscape.

Hickling Broad: a lack of common ground

Perhaps the best example that I came across of the how a "lack of common sense" can be deployed in Broadland is the case of Hickling Broad in the Thurne Valley. Hickling Broad is the largest of the broads, the location of a National Nature Reserve in the care of Norfolk Wildlife Trust, and a popular spot for boating. In the late 1990s a major dispute erupted, in a sequence of events recounted to me by Meg Amsden, a local artist and environmental educator, who was contracted by the Broads Authority at the time to produce a puppet show on the affair.[1] I was also deeply fortunate to interview Dr Martin George, an expert on the natural history of Broadland, whose research on the history of Hickling Nature Reserve is foundational to my analysis below (George 2016).

In the twentieth century Hickling Broad's waters were heavily eutrophicated, although this was only partly due to agriculture. Instead, the major source of nutrients was a large flock of black-headed gulls (*Chroicocephalus ridibundus*) that roosted on the broad, their population being kept artificially high by a local rubbish tip, which provided them with a large and consistent supply of food (ibid., 51). When the tip was closed, the gulls left the broad, and a huge bloom of stonewort (*Chara* sp.), a nationally scarce water plant grew in the water. Although local conservationists were extremely pleased that a rare plant was growing unexpectedly in the polluted water, the local boating community were extremely angry. The mats of thick foliage clogged the engines of their boats. To make matters worse, this bloom came soon after a successful campaign by the Broads Authority promoting a shift from more-powerful diesel engines to less-powerful electric ones to minimise erosion. The new, less-powerful engines couldn't cope with the weeds, and broke down. A series of tense meetings followed, as different user groups argued bitterly about how best to manage the broad. In the end, the stonewort used up the nutrients in the water, and died back – but the controversy had left serious ill-feeling in the local community.

At the time, two academics from the UEA Environmental Sciences department, Tim O'Riordan and Rosie Ferguson, were asked by the Norfolk Wildlife Trust, the Broads Authority, and Natural England to interview the various stakeholders associated with the broad; including employees of the conservation bodies, boaters, and residents. O'Riordan and Ferguson found that, although the stakeholders shared a common vision of Hickling Broad as a place that supports both a healthy ecology and a thriving recreational culture, there was also a significant lack of trust between the main parties, exacerbated by a lack of clarity over management responsibilities, and a tendency for groups to form around common interests, which reinforced extreme and inflexible positions (O'Riordan and Ferguson 2000). Ferguson and O'Riordan argue that the root of the problem – the climate of mistrust – was caused by misperceptions over the intentions of others, a failure to talk face to face, and a lack of appreciation and mutual respect for how each party valued the broad itself. I would add a further layer of commentary here. What we see at Hickling is the playing out of a tacit assumption that underlies any accusation that someone lacks common sense – namely, that anyone who understands a situation *properly* will agree with you about the right and wrong ways to respond to it. In saying "such-and-such lacks common sense" or "this person does one thing, when it would be common sense to do quite another", the speaker is working with the presumption that there exists a sympathetic audience to his remarks who share his own fulsome understanding, as opposed to his interlocutor, who does not. But as we see at Hickling, it is not just a fragmented *understanding* that leads to conflict, but fragmented *interests*. It was in the interests of boaters to have a broad clear of weed, while the conservationists were interested in the opposite outcome. Although everyone may agree on a general vision of what the broad should be like – biodiverse and supporting recreation – when it came to the specifics of how this vision should be realised, different expertise and divergent interests led to differences of opinion. But the assumption that it is a *lack of understanding* that leads to these different managerial positions shifts emphasis away from the diversity of individual expertise, preferences, and needs, towards collective, pragmatic norms. The opposition are transformed from a differently interested party with their own set of expertise (Harvey 2007; Bodenhorn 2014), to fools and incompetents who violate basic norms. This impression is compounded by infrequent, shallow communication – as Ferguson and O'Riordan point out – via email or official letters.

The Hickling case can be compared to the models of common resource management developed by economist Elinor Ostrom. Criticising the prevailing tendency of policymakers to regard public ownership and privatisation of common resources as the only means of protecting those resources, Ostrom argues that alternative institutions – based on the collective management of pooled resources by extractors – can in certain circumstances be highly effective, because they account for the fact that individual choices and behaviour can and do vary according to circumstances (Ostrom 1991, 8–23). Ostrom identifies eight design principles for common-resource management systems that succeed and endure:

> 1 Clear boundaries – that define both who has rights to a resource, and the bounds of the resource itself.
> 2 Rules regarding extraction and management that reflect local conditions.
> 3 A participatory system of management, including all stakeholders.
> 4 Effective monitoring by auditors who are either directly accountable to the stakeholders, or are stakeholders themselves.
> 5 Graduated sanctions, meted out by other stakeholders, or by officials accountable to stakeholders.
> 6 Accessible, low-cost means of conflict resolution.
> 7 Autonomy from external authorities – especially governmental.
> 8 [for common pool resources (CPRs) that are part of larger systems] "Nested enterprises" – CPR bodies existing in multiple layers.
>
> (Ibid., 90)

These measures facilitate a key process in Ostrom's models:

> In many field and experimental situations individuals tend to use heuristics – rules of thumb – that they have learned over time regarding responses that have, in the past, brought them good outcomes in particular kinds of situations. Over the course of frequently encountered, repetitive situations, individuals learn heuristics that are better tailored to the particular situation. With repetition and sufficiently large stakes, individuals may learn heuristics that approach best-response strategies.
>
> (Ostrom 2003, 40)

Ostrom's CPR model attempts to develop these experience-based heuristics into effective tools for resource management; something that formal state regulation cannot achieve. These heuristics can be thought of as applying both to technical practices as well as to social actions; indeed, Ostrom claims, "One might think of norms as heuristics that individuals adopt from a moral perspective in that these are the kinds of actions they wish to follow in living their lives" (ibid., 41). It is clear that, for Ostrom, just as people are better informed about resources they know well, the same is true regarding their interpersonal relationships, with face-to-face interactions being the best method for becoming so informed (Ostrom 2009, 113). The heuristic capacities to which Ostrom is referring here are more or less synonymous with the definition of common sense discussed above in Chapter 2 – a normative and pragmatic reasonableness in a specific working environment, or "taskscape". For Ostrom, then, effective commons management requires common sense. When these processes are absent, Ostrom believes that other outcomes – including the "tragedy of the commons" (Hardin 1968, 1244) are possible:

> When individuals who have high discount rates and little mutual trust act independently, without the capacity to communicate, to enter into binding agreements, and to arrange for monitoring and enforcing mechanisms, they are not likely to choose jointly beneficial strategies unless such strategies happen to be their dominant strategies.
>
> (Ostrom 1991, 183)

But when individuals interact regularly within "a localised physical setting" for an extended period of time, they gain understanding of their shared interests, exchange promises, learn whom to trust, and develop common norms, which in turn support collective management practices (ibid., 184). Quotidian communication, here and elsewhere in Ostrom's work, is thus identified as a key component in the building of trust (Ostrom 2003, 33).

But trust in Ostrom's reading is not absolute – it is highly context-dependent, varying according to particular socio-economic and moral circumstances (Ostrom and Walker 2003, 5). Due to the large number of individuals that rely upon them, the variety of activities that impinge upon them, and the difficulty inherent in their privatisation, water catchments like the broads are prime, if challenging, candidates for management under a CPR format. Ostrom discusses a theoretical example of such a structure in her contribution to *Trust and Reciprocity* (Ostrom 2003). Within the scheme she sketches out, crucial junctions of possibility emerge where the presence or absence of trust plays a key role, such as the "Development of Subgroups" and the "Level of Cooperation" (ibid., 59).

Although Ostrom identifies subgroups as a necessary precursor to converting a watershed – or any resource involving a large variety of heterogeneous users – into a CPR, the Hickling case above indicates that the "Development of Shared Norms" through "Face-to-Face Communications" can actually make the maintenance of cooperation between subgroups more difficult. Ostrom herself presents a solution to this: "nesting" of CPR institutions would be required, beginning with small numbers of landowners operating within the catchment of a single tributary. Face-to-face interactions *within* subgroups should be accompanied by face-to-face interactions *between* them. Ostrom hypothesises that were one such organisation to be created, the economic benefits would encourage other commons to be created in other tributaries; these would then be able to work together to better manage the entire catchment, through face-to-face interactions among appointed representatives.[2] With this proposed solution – that the commoning of a large system must consist of smaller, constituent commons – it is important to note that the boundaries of these smaller commons are defined by the resource itself, namely, the catchments of tributaries.

Ostrom's modelling of common pool resources – stressing the reasonableness of heuristic strategies and conversational norms, taking place within clearly defined commons – is an academic articulation of many of the cultural attitudes I witnessed in the field.[3] But as I have already indicated, the vituperative disagreements at Hickling demonstrate how the hopeful potential of Ostrom's models can be frustrated – not least by the formation of subgroups with divergent interests who fail to speak to one another face-to-face, creating a climate of mistrust.

Although there are common rights at issue on Hickling Broad itself – namely the right of navigation, as well as some fishing rights – these constitute a fragment of what was once a far more extensive network of common rights and obligations within the parish. As we have already seen in Chapter 3, the process of enclosure resulted in a profound social upheaval across England, in which many rural people were made strangers in their own parishes – dislocated from their "localised physical setting", to use Ostrom's phrasing. Regardless of whether the medieval peasant economy actually fostered the development of Ostrom-style heuristics, these heuristics and networks of face-to-face intimacy are hard to sustain in the marketised landscape of progressive enclosure, where private land ownership and possessive individualism prevail. The parish of Hickling is a case in point, with huge areas of common fen and marsh being enclosed and drained in the Enclosure Awards of 1808 (George 2016, 22–25), leaving only two tiny parcels of land for the use of the poor. In the centuries over which enclosure took place, social structures where communal land-use played a key role were progressively replaced by one in which private ownership – and the possessive individual – took centre stage. Williamson observes that

> traditionally the common fens had been used in a variety of ways, and careful management was necessary in order to prevent conflict and argument. After enclosure, however, areas of private fen could be used by their owners in whatever way they chose.
>
> (Williamson 1997, 103)

As in the rest of England, enclosure diminishes the incentive for working together at Hickling. Furthermore, those few fragments of common land that remained in the nineteenth century – the poor allotments – were, unlike the old commons, managed exclusively by middle-class residents; not the actual poor people who used them (ibid., 99). Prior to Parliamentary Enclosure, the wet commons of Broadland were managed by manorial courts. Although such institutions were not independent of elite control, commoners nonetheless had a direct say in how the commons in the parish were managed – unlike the poor allotments (ibid., 78) (see Chapter 3).

Hickling Broad is privately owned – 644 acres of the site were recently purchased by Norfolk Wildlife Trust after a major funding appeal (BBC News 2017; Norfolk Wildlife Trust 2017). The merits of the Trust's management of the site aside, such an arrangement can be compared to that of the poor trusts, in which land is managed privately by a small group on behalf of a particular interest (that of the poor; that of conservation), rather than collectively by all the residents of the parish or manor. This pattern of rights and responsibilities being segmented according to interest, I suggest, is not confined to conservation management for public benefit, but could equally be applied to all the distinct user groups that emerged in the course of the Hickling controversy – from fishermen, to farmers, to boaters. And this segmentation is accentuated by the predominance of private property rights over common ones.

Ultimately, what this piece of historical context serves to demonstrate is the extent to which the socio-economic common ground at Hickling – materialised by the common marshes and fens – has been undermined by an ongoing process of enclosure. Although we saw, in Chapters 2 and 3, how common understandings could emerge among land managers, in society at large possessive individualism is hegemonic. C. B. Macpherson suggests that, in Britain and other modern liberal democracies, "Society consists of relations of exchange between proprietors" (Macpherson 1962, 3), identifying this as a key assumption of what he calls "possessive individualism" (ibid., 263–264), an ideology that also reifies the autonomy of the individual from any kind of prior social relations, a right that is protected behind a veil of legal obscurity (Frake 1996, 104). I argue that the Hickling case indicates that, rather than individual interest and the common being held in balance as land managers strive to achieve, the individual – and the possessive individual at that – is the dominant economic actor in Broadland society, rather than the commoner. As we see with the privatisation of the UK water supply, whether she is dubbed a citizen or a customer, the possessive individual, unlike the commoner, is not directly responsible for the management of their resources, and can only engage those who are in limited ways (Page and Bakker 2005, 54). One businessman felt the implications were stark:

> In order for us to have a viable economy, we actually have to tear apart society … just look around you … it doesn't matter if this is urban or

> rural – what we're actually creating is a completely them-and-us society, and a them-and-us society means we hate government, government hates them, we hold everybody in contempt, everybody is trying to screw us for as much money, and ultimately it doesn't work.

The optimistic assumptions of Ostrom are systematically undermined by a hegemonic state of enclosure where interested, possessive individuals struggle over what few commons are left – creating the very fragmented society I found so methodologically challenging. The segmentation this creates, even within geographically distinguishable commons like the catchment of Hickling Broad, engenders a "them-and-us society" of which the Hickling Broad controversy is a prime example. Deprived of a shared working environment, people increasingly form subgroups based on individual interest and expertise, rather than shared resources. Subjected to the continuing process of enclosure that makes working heuristics and face-to-face relationships difficult to sustain, acrimony and suspicion – often based around entrenched subgroups – was a likely outcome (Rothstein 2000) (see Figure 5.1).

Bird farmers: Catfield Fen and landscape-scale conservation

I first visited Catfield Fen as part of my volunteering with the Bure Valley Conservation Group, a community conservation society set up in 2015 with

Figure 5.1 Continuing enclosure. The cricket ground in the centre of the village of Brundall was sold off for development during my fieldwork – replacing common open space with private properties.

the assistance of Norfolk Wildlife Trust. Consisting of roughly 30 largely retired local people, and with the help of two staff members from Norfolk Wildlife Trust, the group would meet every Wednesday morning at 9.30am outside Acle Library, before heading out to conduct practical conservation work: coppicing trees, planting saplings, raking wildflower meadows, or building bird boxes. The group would also periodically hold talks on naturalism or conduct site visits to places of ecological significance in the Bure and Ant Valleys, after our work for the day was done. One misty January morning, it was announced by that we were in for a special treat – we were going to visit Catfield Fen.

After a short drive, past the little village of Ludham and on, across the wide fields beside the Ant, we parked up under a long stand of wintry oak trees, piled out of the minibus, and headed off down the footpath that ran adjacent to the reserve. As we turned down the Rond – a raised grassy bank that bisected that part of the reserve – the flat landscape stretched out around us in foggy silence, the reeds a sheet of beaten copper under a silvery sky. We were told about the place, how it was officially designated as the finest valley fen in Western Europe, how it was inhabited by otters (*Lutra lutra*) and water rails (*Rallus aquaticus*) and great diving beetles (*Dytiscus marginalis*). As I looked into the murk, I imagined that somewhere out there, red deer (*Cervus elaphus*) would be looking back at us – Britain's largest wild mammal, synonymous in my mind with the wild. Most important of all, Catfield Fen was home to some 5,000 fen orchids (*Liparis loeselii loeselii*) – some 90 per cent of the entire UK population. Part of what made it so special, and so suitable for the orchids, we were told, was that it had been sealed off from the main Broads river system, preserving the quality of its spring-fed waters. Catfield, then, seemed like a magical place – tucked away from the problems of eutrophication and under-management that dogged other parts of Broadland.

Catfield Fen covers some 57 acres of open water and fenland, adjoining some 58 acres of carr woodland, lying to the east of the River Ant in the North Rivers region of the Broads (Harris *et al.* 2008). Half of Catfield Fen is part of the Catfield Estate, and is privately owned by the Harris family, who purchased the site in 1995 following the death of its former owner. The remainder of the fen is split into two nature reserves. Catfield Fen Nature Reserve, owned by Butterfly Conservation, was purchased by their Norfolk Branch in 1992 as a stronghold for the nationally scarce Swallowtail butterfly; Britain's largest. The neighbouring Sutton Fen is owned by the RSPB, who also manage Catfield Fen Nature Reserve on behalf of Butterfly Conservation.

Neither Catfield Fen Nature Reserve nor RSPB Sutton Fen is open to the public, due to their ecological importance. It is for this reason that, when I started as a residential volunteer at Strumpshaw Fen, my boss O insisted that I take the opportunity to visit Sutton Fen. As I found out more about Sutton, I realised that biodiversity wasn't the only reason I had to visit the site: the whole of Catfield Fen was at the centre of a dispute over water abstraction.

Far from my initial impression of romantic wilderness, in truth Catfield was as thoroughly enmeshed in controversies over environmental management and stewardship as any part of the Broadland landscape. After a couple of weeks of emailing back and forth to determine a day when we could be hosted there, we decided upon 8 September – a work day for the trusted team of volunteers who helped the RSPB wardens manage the site. G – one of the other residential volunteers – and I drove away from Strumpshaw, leaving the Yare Valley behind and headed north. The weather was warm and summery, and the roads were clear. We parked up beside the barns where the reserve's equipment was stored and the fields of grain that surrounded the wetland stopped, and got out of the van. As G headed off to chat to the other volunteers, the Site Manager E led me into the site office to pick out some maps. E mentioned that he'd put the folder on water abstraction to one side and hadn't looked at it for quite some time, although the dispute was rumbling on. E also mentioned that the matter was being taken to Public Inquiry – at which he was giving evidence. The Inquiry would be led by a planning inspector, who would make a recommendation to Liz Truss, the Secretary of State for the Environment, who would make the final decision. E was keen to ensure that his evidence was as robust and comprehensive as possible. Once we'd selected a couple of maps from a variety of folders – the one on water abstraction was by far the largest – and we headed out into the fen.

The very structure of the landscape reflects continuing management that responds to broader fluctuations in the market. Parts of the RSPB-managed site had been dug for peat during the eighteenth century on a small scale, the peat being used locally as fuel. The patchwork of dug and undug areas creates the range of habitats we saw out on the fen, because dug and undug peat succeeds at different rates – dug peat succeeds quicker, being more porous and more readily colonised by trees and shrubs. Peat cutting for fuel had now ceased, however. The fen was also cut commercially for marsh hay and litter up until the 1940s. Commercial reed growing for thatching continued until 2010 when the growing acidification and drying of the fen diminished the quality of the reed to such an extent that it wasn't commercially viable. This is unusual in the Broads, as normally the management of the fen by having the reed cut and taken off would prevent litter from building up and would keep the ground level (and thus the pH) relatively consistent. But these traditional practices were no longer effective. Although it had been proposed that the sluice, sealing the fen off from the River Ant, could be opened to allow in an influx of more alkaline water, this would bring salt, pollutants, and fertiliser with it. As such, opening the sluice was a last resort. As we crossed a ditch, E said that the water in the ditches was still very alkaline, because they were deep enough to cut down into the alkali layer. So why was the upper soil so acidic?

The acidity of the soil was a salient factor in the dispute over water abstraction. Water had been abstracted adjacent to the fen since 1986, for use in local agriculture (Harper 2015). In 2008, the Norfolk and Norwich Naturalists made a survey of the fen and concluded that it was drying out due to this abstraction. These concerns were passed on to Natural England and the Environment Agency, the two statutory bodies with responsibility for water abstraction and the maintenance of protected habitats like Catfield Fen (Harper 2014). On 8 May 2015, the Environment Agency denied the application by a local farmer – Q – for two water abstraction licenses that he used to irrigate his fields to be renewed, a decision which Q appealed. A coalition of environmental organisations – including Natural England, Butterfly Conservation, and the RSPB, provided evidence in support of the Environment Agency's original decision (Harper 2016). E, as the warden, had a key role in this process – providing evidence that contributed to the Environment Agency's position, and this was used to defend their judgment as the case ascended through the British legal system via progressive appeals. What the RSPB was arguing, E explained, was that the abstraction of alkaline water from the aquifer that fed the fen was causing the water level in the fen to drop. As the fen dried out, in certain places acidic rainwater was collecting. Without sufficient alkaline water from the aquifer to dilute this rainfall, the soil in these patches became more acidic, which created a foothold for *Sphagnum* moss. This moss had the ability to further acidify the soil, allowing it to spread across the surface of the fen from these footholds, killing alkali-loving plants like the fen orchid. E showed me one map – little yellow dots each showed a fen orchid, while irregular blobs of red showed *Sphagnum* patches. Pointing at one large red patch in the southern corner of the reserve, E observed: "There used to be a good number of fen orchids there. Now they're gone." This information was submitted to a public inquiry regarding Q's appeals, which eventually ruled against further abstraction (Case 2016).

E said that "conservation is as much about people and relationships, as it is about ecology." Engaging with diverse stakeholders and working with people was a vital part of his role, and wasn't something falling within his comfort zone: "I know how to cut back scrub, cut reed, survey fen orchid, keep cattle, and build fences … but dealing with people is a whole different exercise." This more social dimension to his role was a relatively recent development in the lifetime of the RSPB as an organisation. E mentioned how in the past, the RSPB only managed their sites with wildlife in mind, but this was changing as the RSPB realised the importance of working with people. From neighbouring landowners to visitors, the staff of the RSPB were being strongly encouraged to see engaging others as vital for improving outcomes for wildlife. As part of a growing awareness in the conservation community that the whole of society's relationship with the natural world needs to be changed if biodiversity is to be protected, the RSPB has focused increasingly upon engaging other groups in society as part of its Saving Nature Strategy (Clarke *et al.* 2013). In the Broads, this strategic imperative was reflected by two principal initiatives: *Futurescapes* and *Networks for Nature.* While *Networks for Nature* seeks to build engagement among the public – being directed toward changing gardening practices to create habitats – *Futurescapes* seeks to build a common methodological approach among land managers in the region that encourages wildlife. Initiated in 2006, *Futurescapes* represents the RSPB's national contribution to landscape scale conservation – "the coordinated conservation and management of habitats for a range of species across a large natural area, often made up of a network of sites" (Bourn and Bulman 2005; Ellis *et al.* 2011, 1).

As H, another warden at Strumpshaw, explained – isolated nature reserves come with a series of problems. They only represent a very small part of the total land area, and they concentrate populations of rare species – a single extreme weather event or predator can have a dramatic impact on the overall population, as the situation at Catfield Fen demonstrates. Landscape-scale conservation addresses these issues, by extending conservation practices into the wider landscape. The Broads catchment is the leading Futurescape in the UK and is therefore treated as an example of best practice within the RSPB. One part of this initiative was Broads Land Management Services, through which the RSPB made its skills and specialist equipment available to landowners through contracting. When the 540 agri-environment scheme agreements in the Broads were concluded in favour of Entry Level Stewardship and Higher Level Stewardship agreements in 2005, the RSPB moved to advise landowners on how to become eligible for the new funding sources:

> It's good for us, but also good for the farmers. Some aren't just into growing cows, they're also really into their nature. It's not too dissimilar to our ANEs [Active Nature Enthusiasts – a segmentation of the public developed by the RSPB as part of *Networks for Nature*] – you just need to find an "AFE"....

Tellingly, V concluded: "The only way you can do that is through personal connections, knocking on doors...."

Futurescapes served to build a respectful, collaborative relationship between the RSPB and other land managers in the region, including farmers. Indeed, one of my informants referred to the RSPB archly as "bird farmers". But V embraced this epithet:

> A landowner owns some arable land, he plants a crop, he's invested a lot of money in that land, so why should his crop fail? The RSPB has invested a lot of money in an area of land to grow a crop – birds, plants, invertebrates etc. – why should that suffer?

For V, this opinion encompassed the fact that conservationists and farmers – despite sometimes being at loggerheads, had a significant amount in common. "We have our own agronomists; they're not called agronomists, they're called ecologists!"

There are several threads here that are insightful, in comparison to the Hickling case, and to what a "lack of common sense" reveals about the controversies surrounding land management in the Broads. The previous, "reserve-centred" approach to conservation – where privately owned tracts of land were stewarded for the public benefit, in this case to conserve wildlife – shows continuity with the management of poor allotments, involving a professional body maintaining a small amount of land for wider public benefit, within a largely privatised landscape. But such provisions do little to change the character of the overall landscape and the adverse consequences this has on wildlife. Wildlife, after all, does not respect property boundaries. Possessive individuals, in complete control of their private estate, are free to pursue practices in their interests that damage the public good. The biodiversity crisis in the UK today demonstrates this in action. Having recognised this problem, the RSPB, like other wildlife charities in the region,[4] have refocused some of their efforts upon building face-to-face contact with their neighbours. *Futurescapes* indicates that a key component of this approach has been emphasising the *continuity of interest* between the RSPB and other land managers – wild species simply being a different "crop" to domestic ones. Although disputes may emerge in such a context, one of the two kinds of framing with which we began this chapter – of anonymous opponents who lack common sense, versus the known individuals who possess it – was conspicuously absent from this initiative.

Fragmenting corporeal attitudes: habitus and "the silo effect"

This series of ethnographic cases of Broadland disagreements, with or without the framing of scarce common sense, invites consideration of an analytical approach that has extensive relevance: from the micro-political disputes of the Broads, to a macro-political one with direct implications for Broadland – namely, the 2016 UK referendum on membership of the European Union. This approach – drawn from a popular reading of the social theory of Pierre Bourdieu – will be dealt with in depth below.

On 1 September 2015, while I was working on the fields and marshes of Strumpshaw Fen, Gillian Tett, a British financial journalist, anthropologist, and author, published *The Silo Effect: The Peril of Expertise and the Promise of Breaking Down Barriers* – a book that Steven Poole of the *Guardian* dubbed a "subversive manifesto" (Poole 2015). In *The Silo Effect*, Tett argues the following about complex modern institutions, from multinational corporations like Sony to state bureaucracies like that of New York City:

> Many large organisations are divided, and then subdivided into numerous different departments, which often fail to talk to each other – let alone collaborate. People often live in separate mental and social "ghettos", talking and coexisting only with people like us. In many countries,

politics is polarized. Professions seem increasingly specialized, partly because technology keeps becoming more complex and sophisticated, and is only understood by a tiny pool of experts.

(Tett 2016, 16)

These highly inward-looking and exclusive subdivisons, Tett claims, are structural features of contemporary cultural life, manifesting both in social organisation and in terms of the categories we use to make sense of the world. Tett dubs this structural fragmentation "siloing", and suggests that it leads to intergroup conflict, tunnel vision, and a blindness to both risk and opportunity. As Green puts it in her review of Tett's book, "Closed, self-referential networks where socially constructed truths prevail and established ways of doing things are never challenged amount to silos which stifle innovation, limit adaptiveness and lead to organisational failure" (Green 2016). What is so "subversive" about this analysis of modern corporate life is that it runs counter to the prevailing assumptions about bureaucracy that presume greater streamlining and specialisation in the workplace lead to efficiency and reliability (see du Gay 2000; Woolley 2018, 168–197).[5] The cases marshalled by Tett – of both the success of those who master their silos and the failure of those who are mastered by them – demonstrate that this is not always the case.

Tett's *The Silo Effect* uses a clear and accessible parse of the life and theoretical contributions of Pierre Bourdieu – suitable for her wide audience – to develop her analysis. The silo mentality that fragments large organisations, and blinkers vision, is a consequence of the tendency of human beings to naturalise their own habitus – a tendency that Tett exhorts her readers to overcome (Tett 2016, 61). Tett doesn't portray silos as necessarily bad – acquiring a specific habitus being part and parcel of social life – but the critical issue is whether people are able to step *outside* of those structuring concerns. Those who are able to mentally or physically disrupt the structures to which they are used, are able to staunch the problems that siloing creates. Those who do not, face potentially serious consequences.

As with her predictions in 2006 that a financial crisis was imminent (MacKenzie 2009; McKenna 2011), Tett's warning concerning the dangers of siloing proved prophetic. Less than a year after *The Silo Effect* was published, the electorate of the United Kingdom voted to leave the European Union, a result that shocked the intellectual and political establishment and sent an earthquake through the currency markets (Wheeler and Hunt 2016). The vote in favour of Brexit suggests that both of the silo effects described by Tett – namely, the isolation of different groups within a common organisation and the mental inflexibility this engenders – were at play in British politics. On the one hand, the result is deemed to have revealed a lack of faith in the "liberal elite" that fuels a populist mood (Münchau 2017), revealing a major divide between those who believed numerous expert voices warning of the dire consequences of a Leave victory – such as the IMF, world leaders, and the UK government itself – and those who distrusted those voices (Foster

2016; Van Reenen 2016). Second, few in the British establishment saw the result coming (Parker *et al.* 2016), with Prime Minister David Cameron resigning in the aftermath, as many commentators suggested his judgement of the public mood and his assessment of the likelihood of a Remain victory was fundamentally flawed (Kershaw *et al.* 2016). The country appeared to have been split into two enormous silos, leaving it totally unprepared for the "shock" of the Brexit vote.

What struck me most about the discourse surrounding the vote was the extent to which common sense was invoked regularly in media commentary. For the Remain side, Brexit meant economic ruin and destroyed our previously cordial relationship with European nations – as such, it flew in the face of common sense (Goet 2016). Laurie Penny, for example, laments that

> there's not enough tea in the entire nation to help us Keep Calm and Carry On today. Not on a day when prejudice, propaganda, naked xenophobia and callous fear-mongering have won out over the common sense we British like to pride ourselves on
>
> (Penny 2016),

while Zoe Williams exhorts us to "insist upon a snap general election as a matter of common sense" as a practical way to fix the crisis of Brexit (Williams 2016). To those who supported Leave, however, the result was a victory for common sense – a reassertion by sensible folk of control over matters that had been delegated to distant bureaucrats in Brussels and a metropolitan elite, whose "expert" decisions regularly frustrated common sense (Dunford and Kirk 2016). From this point of view, as Charles Gasparino put it, "Brexit won because common sense prevailed" (Gasparino 2016). Common sense itself, so it appears, had been cleaved in two. Consequently, the issue of whether to leave or remain in the European Union revealed deep differences between how large groups of people in England understood the world and chose to act within it.

The Broads was by no means removed from this national political context. Martin Harper, the RSPB's Conservation Director, greeted the Catfield Fen decision in favour of the Environment Agency's abstraction ban with the following observation: "While there has been a bit of a reaction from the farming community about this, this is just another example of the big conversation that we need to have as a result of the Brexit vote" (Harper 2016). Conversations aside, leaving the European Union would have a direct impact on the Broads in two main respects. First, the EU has a broad framework of environmental protections – such as the Habitats Directive and the Water Framework Directive – that would cease to apply in Britain once the United Kingdom left, unless transcribed into UK law (Neal 2016). Second, the Common Agricultural Policy (CAP) provides direct subsidy to British farmers and landowners, amounting to some €3,084 billion in Pillar 1 payments in 2015, and access to a further €5.2 billion in Pillar 2 (rural development) funding over the same period (the National Farmers' Union 2015). The management of much of Broadland is supported by agri-environment Pillar 2 subsidies – such as NELMS (the old scheme) and Countryside Stewardship (launched in 2014), both of which provided financial incentives to keep farmers from ploughing up and deep-draining marsh and fen. Britain's departure from the European Union would mean that these financial and legal arrangements will be subject to change. The big conversations around Brexit, and the direct impact Brexit would have on Broadland, prompted me to consider carefully how Tett's notion of "siloing" – the blinkered vision that results from people allowing themselves to be limited by their specific habitus – might represent a key process within Broadland social life, and English society more generally.

Tett's interpretation of Bourdieu draws extensively on one of his more ethnographic texts, *The Bachelor's Ball: The Crisis of Peasant Society in Béarn* (Bourdieu 2008). Here, Bourdieu turned his critical attention to a cultural milieu with which he was intimately familiar: the rural region of Béarn, where he grew up. He centred his ethnography around a vivid account – quoted by Tett (2016, 25–51) – of how at dances, the young women and fashionably dressed men jiving in the limelight were surrounded by a corona of men in their 30s, dressed unfashionably, standing still and looking on (Bourdieu 2008, 81–83). These poor souls were deemed by local people to be "unmarriageable"; Bourdieu set out to explain why. Bourdieu embarked on a sophisticated description of Béarnaise kinship practices, detailing how the practice of primogeniture, combined with the payment of dowries to younger siblings once ensured the continuity of each farming estate was not threatened by the contradictory ambition to ensure equity of inheritance among siblings (ibid., 38). However, due to intersecting economic trends following the Second World War, together with the penetration of modern forms of symbolic capital into rural communities, this same arrangement put the elder sons of peasant families who stood to inherit in a disadvantageous position relative to their younger brothers, or men from elsewhere, who worked in the towns and villages. As such:

> The norms governing the selection of a partner were valid in rough terms at least, for the whole community: the accomplished man had to combine the qualities of the good peasant and the sociable man, and find a proper balance between lou moussu and lou hucou, the rustic and the urbanite. Today's society is dominated by divergent systems of values: alongside the essentially rural values that have been defined, new values are emerging, borrowed from the urban world and adopted mainly by the women; within this logic, priority is given to the "monsieur", and to the ideal of urban sociability, quite different from the old ideal, which was mainly directed towards relations between men. Judged by these criteria, the peasant becomes the hucou.
>
> (Ibid., 46–47)

Peasant heirs thus became "empeasanted". The village, once integrated with the peasant farms that surrounded it, became culturally antithetical to them:

> [Thus] the barrier between town and country, between the peasant and the townsman, which used to run between people from Pau and Oloron and the people of Lesquire, without distinction, now separates the villagers, lous carrerens, from the peasants of the hameaux. The opposition between peasant and townsman now starts in the very heart of the village community.
>
> (Ibid., 71)

Consequently, the

> small country ball is the scene of a real clash of civilisations. Through it, the whole urban world, with its cultural models, its music, its dances, its techniques for the use of the body, bursts into peasant life. The traditional patterns of festive behaviour have been lost or have given way to urban patterns.
>
> (Ibid., 83)

The desires of young women, prefigured by a shift in the relative economic station of the peasantry, are out of keeping with the modes of dress and bodily hexis of its heirs (ibid., 81–93). Overall, Bourdieu summarises his project as follows:

> My intention is simply to evoke the set of processes which, in the economic order but also and especially in the symbolic order, have accompanied the objective and subjective opening of the peasant world (and, more generally, the rural world), progressively neutralizing the efficacy of the factors which tended to ensure the relative autonomy of that world and to make possible … a form of cultural particularism, based on a more or less self-confident resistance to urban norms particularly as regards language, and a kind of localocentrism, in religion and politics.
>
> (Ibid., 183)

The erosion of this "localocentrism" – what we might call "parochialism" – due to the economic and social changes to French society in the course of Bourdieu's lifetime, had ensured that "this closed world in which people felt themselves to be among like-minded people has gradually opened up" (ibid., 183). In short, the Béarnaise bachelors were victims of a historical process, in which the social capital of their habitus became dramatically devalued, being on the wrong side of a widening gulf between urban and rustic French society. To recall Wolf's circles of decreasing familiarity, mentioned in Chapter 3, the "unmarriageable" peasants of Béarn were overcome by the circles of the merchant and other outsiders.

Though the historical circumstances are quite different, Bourdieu's work here gives us a sense of how particular kinds of what Bourdieu calls "corporeal attitudes" – i.e. habitus (ibid., 84, 134) – can reinforce social divides that can constrain the rational strategising of those on either side. This is highly pertinent to the Hickling case, and helps us to understand how Ostrom-like commons management failed to form there. In a landscape that is heavily privatised, people's social expectations are underpinned by a prevailing assumption of possessive individualism, i.e. that each individual can (and should) pursue their interests independently from society – with society itself therefore consisting of "relations of exchange between proprietors" (Macpherson 1962, 3). "Political society" meanwhile, "is a human contrivance for the protection of the individual's property in his person and goods, and (therefore) for the maintenance of orderly relations of exchange between individuals regarded as proprietors of themselves," where individual freedom is experienced most keenly through property relations (ibid., 263–264). As Macpherson goes on to point out, this system requires that all those with a voice in choosing the government must have basically coherent interests, a condition that was previously met by restricting suffrage to the propertied classes, but which broke down under universal suffrage (ibid., 273). We see this same process – by which the political order built upon possessive individualism is destabilised by diverse interests – happening in microcosm at Hickling. Though groups and associations of Hickling today do exist, they form around shared interests – fishing clubs, professions, boating companies, conservation charities, and local residents associations. Such groups have become socially and practically isolated from one another, something that leads to the development of their own distinctive habitus. Rather than the commons of Hickling being managed by all the individuals in the parish collectively, each interest group becomes, in a sense, a parish in its own right, distinguished by particular practices, possessions, styles of dress, and environmental management priorities that immediately mark them out, being particular to their interest – discussed with respect to the visitors to Strumpshaw Fen in Chapter 4. However, unlike the classic model of the parish or manor described in Chapter 3 – in which a group of human individuals share exclusive rights to a defined area of land – these interest groups today have overlapping claims to the same tracts of country (usufruct of the broad) as well as a core of private property they own absolutely (possessions, boats, private homes, etc.) – either as individuals, or as an incorporated association, which is itself legally an individual. As these overlapping claims often conflict with one another, political society at Hickling – that under possessive individualism should guarantee private interests – instead is perceived as a *frustration* to those interests. This experience is prefigured by a habitus of enclosure; day-to-day, the inhabitants of Hickling parish move through a landscape of private property, of mine and yours, of "them-and-us". Only rarely do they encounter common resources; and when they do, the political controversies about how they should be managed are informed by that same habitus of enclosure – engendering a prevailing condition of possessive individualism, with all the conflicts that entails. The fact that possessive individualism has become so pronounced in Broadland society means that there is little common ground within Hickling that includes everyone, and so the fragmentation of corporeal attitudes and environmental managerial approaches – siloing, in other words – is the natural result. This fragmentation has had concrete physical effects upon the landscape – from the improvement of agricultural land, to declining biodiversity.

The silo – an industrial storehouse for grain, built of metal or concrete, designed to isolate the bulk produce from its surroundings – is a particularly apt metaphor for describing the social conditions that have led to the segmentation of habitus that frustrates common sense; it is a synecdoche for the socio-economic condition of English rural social history. The Broads today has plenty of them, both metaphorical and physical. Perhaps the most striking example of these are the silos of Cantley Sugar Factory, an industrial plant on the edge of the little village of Cantley in the Yare Valley – at the far south-eastern tip of the Mid Yare NNR. For most of my fieldwork, these grey behemoths loomed in the distance, a constant presence in the landscape. Built in 1910 to refine beet – grown locally in vast quantities – into sugar, the factory is just one sign of the presence of the market in the Broads, one set of silos among many. Just as Sydney Mintz used the single commodity of sugar to explore the history of Western colonialism and industrial modernity (Mintz 1986), so I suggest that the silo is an excellent expression of the prevailing condition – one of progressive enclosure – that possessive individualism has created in the Broads. Silos contain the produce of industrial farming, and are a physical sign of agricultural maximisation, just as they capture metaphorically the social conditions of individuation that now prevail in Broadland. Such social conditions are themselves actualised and affirmed by the creation of hedges, fences, signage, and gates – all of which limit access, demarcate property, and instil enclosure by fragmenting the bodily attitudes of local people (see Figure 5.2). Distinct forms of habitus do not necessarily lead to tension, however. As we have seen from Bourdieu's ethnography above, the result of the peasantry becoming siloed was their gradual economic and social marginalisation, rather than the sort of hostility I have described in Broadland. For conflict to occur, a specific set of cultural dispositions is necessary – and this is where common sense, or the lack of it, comes into play.

Trials and errors: the trouble with common sense

To understand how the assumption of common sense, combined with distinct forms of habitus, led an event as dramatic as Brexit and the controversies of Hickling, it is helpful to consider Bourdieu's own discussion of common sense. Bourdieu uses two terms that are often translated as "common sense": *sens commun*, and *bon sens* (see Chapter 1). A number of social scientists who comment on Bourdieu's work have used "common sense" as merely the English language equivalent of the sorts of culturally specific, informally

Figure 5.2 An instance of enclosure.

acquired assumptions that constitute doxa (e.g. Shiach 1993; Hamel 1997; Holton 2000; Wacquant 2001). Bourdieu himself writes, in *Masculine Domination* (2001), that common sense (*sens commun*) is "a practical, doxic consensus on the sense of practices" (Bourdieu 2001, 33). Habitus plays a key role in the perpetuation of this consensus:

> One of the fundamental effects of the orchestration of habitus is *the production of a commonsense world endowed with the objectivity secured by consensus on the meaning (sens) of practices and the world*, in other words the harmonization of agents' experiences and the continuous reinforcement that each of them receives from the expression, individual or collective (in festivals, for example), improvised or programmed (commonplaces, sayings), of similar or identical experiences.
>
> (Bourdieu 1977, 80, emphasis mine)

This "commonsense world" has obduracy because it resides in "the universe of the undiscussed" – that is, doxa.

> Because the subjective necessity and self-evidence of the commonsense world are validated by the objective consensus on the sense of the world, what is essential goes without saying because it comes without saying: the tradition is silent, not least about itself as a tradition; customary law is

> content to enumerate specific applications of principles which remain implicit and unformulated, because unquestioned.
>
> (Ibid., 167)

Common sense-as-doxa is counterpoised to the "universe of discourse", which can be divided into that which is heterodox (that which runs contrary to doxa) and that which is orthodox (attempts to restore doxa, without ever managing to do so) (ibid., 164).

Bourdieu also provides a framework for how social critique and cultural change can develop, and how the doxic "universe of the undisputed" can be brought into dispute. For this to occur, Bourdieu claims, either "cultural contact" or "objective social crises" – such as ecological disasters, political unrest, or economic collapse – are needed:

> The critique which brings the undiscussed into discussion, the unformulated into formulation, has as the condition of its possibility objective crisis, which, in breaking the immediate fit between the subjective structures and the objective structures, destroys self-evidence practically. It is when the social world loses its character as a natural phenomenon that the question of the natural or conventional character (phusei or nomo) of social facts can be raised.
>
> (Ibid., 168–169)

It is crisis, for Bourdieu, that has the power to unsettle our comfortable certainties (Argyrou 2013, 96; also Lizardo and Strand 2010; Morrin 2015). In the gap that crises create between the subjective and the objective features of experience, Bourdieu argues, social critique – perhaps fuelled by experiences of a different cultural context – can be constructed (Bourdieu and Wacquant 2002). Although doxa materialise through structures of habitus and bodily hexis, the universe of the undisputed isn't rigid or inescapable, but is constantly changing according to the experiences of those within it – an "open system of dispositions that is constantly subjected to experiences, and therefore constantly affected by them in a way that either reinforces or modifies its structures" (Bourdieu and Wacquant 2002, 133; García and Spencer 2014, 47). In *The Silo Effect*, Tett clearly draws upon this notion of heterodoxy as her prescription for avoiding siloing. By seeking out different perspectives found in other cultural environs, Tett argues, silos can be busted – both conceptually and socially. Jeremy Lane, by contrast, alleges that "only those with the time and money to stand back from the realm of material necessity could stand back from and achieve critical distance on their own social universe" (2006, 59). Hamel claims that: "The epistemological rupture [of reflexive sociology] was thus marked by an opposition to the actors' practical consciousness which is conveyed by common sense, seen by Bourdieu as false consciousness" (Hamel 1997, 102–103). Hamel's reading of Bourdieu reflects the characterisation of common sense made by Vico, centuries earlier, that common sense is simply "judgement without reflection" (Vico 1948, 142; Marková 2016, 49). However, Hamel and Lane's characterisation of Bourdieu neglects the fact that what he proposes is a theory of *practice*, rather than one of abstract reflection. In *Outline*, for example, he stresses that the interest of the subjugated classes is to contest doxic ideas (Bourdieu 1977, 169). Bourdieu's discussion of the conditions of social critique in *Outline* is not definitive, nor does he specify precisely what sorts of crises can trigger critical movements (Bilic 2010, 380–381).

To apply Bourdieu's theoretical framework to my ethnography above, I'd like to return to the observation made at the outset – that the absence of common sense is invoked in two particular ways. When referred to as a *joke*, it normally implies that the speaker (and the listener) possess(es) it, even if everyone else does not. As we have seen, holding common sense – by definition – yields "good" choices, both morally and practically. This "sense" is not a particular body of knowledge, or set of structures, but rather an *attitude that actively seeks out* those "good" choices – one thinks, for example, of the description offered by Paxman of English "empiricism" involving a process of "snuffling around a problem" until a route around it is found (Paxman 2001). If we were to emplace this meaning of common sense – as an emic category – within Bourdieu's model, it might be fair to argue that it proposes a kind of *practical heterodoxy*. This is not as formalised as the more academic reflexivity imagined by Bourdieu in *Outline*; rather, it is a pragmatic attitude towards particular problems, construed as small-scale crises that prompt critique of received wisdom. Through an iterative process of trial and error, in which solutions are sought to specific issues thrown up in the course of working together, individuals can develop a "common-sense attitude" cast as a virtue that should be shared, but sadly is not. Indeed, conversing with a Canadian forester who had visited a number of Broadland estates enabled the immediate identification of "trial and error" as characteristic of the environmental land management practices he found there. This practical heterodoxy, I argue, serves to characterise the sort of common-sense attitude lying at the heart of Ostrom's model – a habitus of commoning that creates opportunities for sensible working together in a common landscape.

But where a lack of common sense is directed as an *allegation* – directed at specific individuals, rather than ironically at society at large – we see a different set of meanings emerge, and ones that are directly pertinent to the controversy at Hickling. Invoking common sense in this latter way is a classic act of orthodoxy as theorised by Bourdieu: shoring up a doxic set of claims by characterising your opponents as wrongheaded, because they second-guess what is merely natural (Bourdieu 1977, 78).

The reason why there is such conflict at Hickling, I suggest, is because these two quite distinct, even contradictory meanings are carried *by the same term*. This encourages people to view the orthodoxy of their particular group (say, fishermen) *as* practical heterodoxy. This simultaneously makes each group both the moral majority, *and* the sensible, critically minded party.

By drawing upon their group's shared experience – of wrestling with weedy propellers, of excitedly tracking the spread of a rare plant – each person involved in the controversy fortifies their particular interests with experiential knowledge. By imagining that their specific ideas about how the broad should and should not be managed are pragmatic and oriented against "mere opinion" in this way, rather than a particular perspective arising from specific interests and expertise, all of which are structured by a habitus proper to their own group, each group is predisposed to view any disagreement from other user groups as both illegitimate and foolish. This in turn exacerbates a lack of trust, and antagonises the situation further. As such, Hickling Broad acts as a case study of how the distinct kinds of habitus can lead to a silo effect. Conservationists, boaters, local residents – all would have moved through and occupied the broad at different times and in different ways, sometimes even working in entirely separate locations (see Chapter 3). This precluded opportunities for face-to-face contact and engendered a quite distinct habitus for each group. When contrasted with Ostrom's work – which, as I have said, represents a sort of academic distillation of the root metaphor of the common – the aetiology of this problem becomes quite clear. Ostrom insists on the importance of face-to-face contact, clear shared norms, and so on, all of which would constitute each common pool resource as subfield, within which practical heterodoxy can emerge. However, what we see at Hickling is what happens when this ideal situation is perturbed by the complexity of modern life, in which people live and work separately from one another. Rather than being the material expression of a subfield responsible for its management ("the common of Hickling"), Hickling Broad has become a site where multiple different fields, each with their own distinct habitus, clash with one another. This has meant that, in the wake of drastic environmental change in the form of a *Chara* bloom, the stakeholders around the common resource of Hickling Broad had developed all the hallmarks of a silo mentality. Instead of a crisis creating an opportunity for reflection, it had the opposite effect, causing people to retreat into particular "interest groups", and to become increasingly suspicious of others with whom they shared the broad. Since the Hickling controversy, the RSPB, like the conservation sector as a whole, have responded to the practical crisis of a crash in biodiversity by reflecting critically on their established techniques – conditions of practical heterodoxy, which in turn has supported them in stepping outside of their respective "silo".

Thus far, I have drawn together conclusions from my previous chapters – about the nature of common sense as an emic category; about the experience of working in a shared environment, punctuated by ordinary crises and conflicting objectives; and about the role of enclosure as a continuing process within the English landscape. These serve to underpin my contestation here, that, in addition to voicing an orthodox opinion, common sense in English contexts can also denote a folk model of a reflexive attitude – a practical heterodoxy – that is deemed to furnish the possessor with the ability to make sensible decisions, set within a common experience, with reference to a mutually apprehended context. Common sense exists in a state of mutual constitution with common ground. Through trial and error, English common sense and common life is deemed to come into being. This much is what is expected of common sense in its positive form. These expectations, however, fall far short of the reality amidst the lakes and rivers of East Norfolk. This, as I have said, is a land of silos, where common sense is thought of as hard to come by. Indeed, the very polysemy of common sense – as "practical ortho/heterodoxy" – is what gives common sense, even in its absence, its enduring power to frustrate efforts to work together.

Conclusion: Chedgrave Common and the apogee of commoning

Attempting to build connections between heterogeneous factions within Broadland society – especially between the Broads Authority and the wider community – has been on the agenda for many years. M, a former Countryside Ranger, characterised her role as that of an intermediary between the Authority and local people. The goal was to ensure that the inhabitants of villages in her "patch" of the Broads – namely, the Yare Valley – both understood and contributed actively to the management of their local area. This was not an easy task, as local people did not always view the Broads Authority positively. M said:

> Often, I'd meet a landowner, and they say "Oh, you lot ...". And [sigh] they'd kind of have this idea that there'd be hundreds of men in grey suits, doing things when – I had to say "Look, there's only me out here now!" They had this false impression that the Broads Authority was massive and overfunded, and full of these grey suited – um – bureaucrats. When nothing was further from the truth, really.

As we shall see in the next chapter, this attitude is indeed influential within the Broads. Although many of the farmers I spoke to attested to the constructive relationship they had with the Broads Authority specifically,[6] bureaucrats in general were cast as the pitiless enemies of common sense. In order to build bridges, therefore, M had to contest this inaccurate view of the Broads Authority as staffed by faceless administrators, lacking in common sense. To do this, M focused on creating what she called "relationships on the ground" – personal, face-to-face relationships with landowners, land managers, and local residents in her "patch". An added complication to M's work was that even the communities themselves in which she worked were often "multi-layered", with different groups of people who often hardly socialised with one another – "There's different communities even within that geographical space, and they all have different, you know, they're not all singing from the same hymnsheet." Through careful diplomacy and nurturing of "relationships on the ground", M was able to build a collaborative network between a wide variety of stakeholders, from parish councils to schools, from landowners to birdwatchers – not only building links between these groups and the Broads Authority, but between these groups themselves.

Unfortunately, in order to save money in the wake of the 2008 Financial Crisis, the Broads Authority amalgamated the roles of Countryside and River Rangers – who were responsible for monitoring navigation on the waterways – shortly after M's departure from her role. The remaining Rangers were required to spend two thirds of their time on navigation-related duties, meaning that much of the labour M once did went undone. M noted with regret that the landscape itself had been directly, physically impacted by these recent changes:

> For example in my area, and I know this is replicated in other river valleys: I was really lucky, I used to look after a local common at Chedgrave, and I used to do a lot of conservation work there, and we had local working groups, everything like that. Spent many a hot day pulling ragwort. I was there not that long ago, and it's gone to rack and ruin because we're not looking after it anymore; that's not part of our remit.... It's not being looked after by anyone cohesive anymore, if you see what I mean, [and] that was one of our in-roads into the community[,] helping look after that.... It's owned by a parish council poors trust.

As with Strumpshaw Fen and the farms discussed above, the goal here for M was to create a particular sort of habitat – in which succession had been held back and an "aesthetic of proximity" developed, one that was possessed by local village communities and the Authority. But M argued that without continued support, the collective social and ecological possibilities unravelled.

The fate of Chedgrave Common, I suggest, maps the social role of enclosure in the Broads as a whole. As a poors trust, Chedgrave Common, like the poor allotments of Hickling, is a tiny fragment of a once extensive area of common land. Once managed and utilised collectively, it could have been a key site of local economic life, but it is now largely abandoned, surrounded by enclosed parcels that are fully integrated into an industrialised market economy. The conditions of late capitalism maintain a social landscape that is fragmented, so once state intervention is removed, public participation falls away – falling far short of the maximal potential modelled by Ostrom. Cultural theorist Morag Shiach claims that Bourdieu's "analyses serve to specify the terms of our enclosure rather than offer us any escape" (Shiach 1993, 219), a reading that is then picked up by Robert Holton, who describes this phenomenon as "enclosure by common sense" (Holton 2000, 90). Not only does Holton's reading neglect the opportunities for critique Bourdieu proposes as "reflexive sociology" – opportunities that, as my discussion of the RSPB's activities indicate, can be pursued by non-sociologists – but, I would suggest, this "enclosure by common sense" of individuals via Bourdieu's approach is precisely back to front. As an ongoing historical process, enclosure does not represent an effacement of the individual, but rather *its wholesale institution* into economic life. As Strang points out:

> In historic patterns of enclosures in Dorset there is clearly a relationship between control over resources, social identity and the empowerment – or disempowerment – of particular groups. At every stage, increasing control over land and water has supported particular elites, while those excluded from such ownership have developed numerous strategies to try to fulfil their social and economic needs, to maintain their connection with the environment and each other. However, it seems that their enforced departure from the land has massively undermined the potential for the construction of localized communities and encouraged a shift to much more individuated forms of identity.
>
> (Strang 2004, 20)

As we have seen above, the enclosure of the commons replaced the manorial collective rights with individuated property relations. Although the poor would have indeed experienced this as a massive curtailment of their personal rights and privileges, the rich and middle-earners would have seen their individual autonomy expanded considerably, a process that, in turn, helped constitute the modern individual – with John Locke's theory of property demonstrating the genesis of both the modern possessive individual and a theory of property that justified the enclosure of the commons (Macpherson 1962; Locke 1988; cf. Tully 1980, 153–154; Wood 1984, 66). As such, it is somewhat ironic to suggest that the individual action is "enclosed" by common sense, when, in actuality, enclosure represents the effacement of a collectively managed and utilised landscape in favour of an individuated and privatised one. This effacement prefigured the development of individualism as we know it today.

The silos that Brexit revealed at the forefront of British political discourse arise from a landscape of deep socio-economic divisions, rooted in what Tsing characterises as the "ruined landscape" of late capitalism (Tsing 2015). These conditions structure, and are structured by, the habitus of enclosure of possessive individuals. And the impact of Brexit upon the enclosed landscape of the Broads is likely to be profound. The Environmental Stewardship Schemes pioneered in the Halvergate Triangle in the 1970s and 1980s and their successors across this fragile landscape have largely been made under the auspices of the Common Agricultural Policy (CAP), and were themselves made necessary by CAP's productivist orientation in the post-war years. The large landowners I spoke to and interviewed in the Yare Valley all relied heavily upon Pillar 2 environmental subsidies;[7] as one remarked to me, the business of owning a large estate is basically a matter of grants and tax avoidance. The main portion of the former consists of EU subsidies. Equally, a great many regulations that protect the environment and restrict or mandate certain industrial and agricultural practices may be altered or subject to question in the wake of Britain's eventual exit from the EU. Leaving the EU as a geopolitical event will no doubt intersect with the sort of local disputes and initiatives concerning land management discussed in this chapter. The crisis of Brexit throws the doxa of the English countryside into question; and attempts to bust the silos – through activities like the RSPB's *Futurescapes* – will no doubt continue in earnest. Such efforts are, as my informants regularly told me, simple common sense.

Notes

1 "The Bloomin' Weed" performed in 2000.
2 Ostrom also argues that external governmental agencies could have both a positive and a negative role to play: overruling design principle 7 above (see p. 175), Ostrom argues, would catastrophically undermine the networks of trust and reciprocity upon which such institutions rely – while offering expertise and financial support could hasten the commoning process.
3 See in particular the quotations from wardens T and H in Chapter 4, and the reasonableness and specificity of "common sense" in the "Attitude Test" used by F at Whitlingham Country Park, and of the "simple common sense" of P's actions with respect to a community agricultural project in Chapter 2.
4 The Norfolk Wildlife Trust have also introduced their own landscape-scale conservation initiative – of which the Acle Conservation Group was part – called *Living Landscapes*.
5 For a more detailed discussion of the Weberian ideal type of bureaucracy, and for academic critiques of this model, see Chapter 5 of Woolley 2018.
6 However, there is a distinct possibility that the farmers said this to me because they thought the Broads Authority would read my findings, and therefore concealed their true feelings in order to avoid controversy.
7 Pillar 2 subsidies are a tranche of funding within the Common Agricultural Policy that is earmarked for rural development. It contains three "axes" – one of which is designed to incentivise landowners to conserve and enhance the conservation value of the rural landscape – see above.

References

Argyrou, Vassos. 2013. *The Gift of European Thought and the Cost of Living*. Oxford; New York, NY: Berghahn Books.

BBC News. 2017. "Norfolk Wildlife Trust Secures Funds to Buy Hickling Broad". *BBC News*, April 18, 2017, sec. Norfolk. www.bbc.co.uk/news/uk-england-norfolk-39633959.

Bilic, Bojan. 2010. "Bourdieu and Social Movements Theories: Some Preliminary Remarks on a Possible Conceptual Cross-Fertilization in the Context of (Post-) Yugoslav Anti-War and Peace Activism". *Sociologija* 52 (4): 377–398. https://doi.org/10.2298/SOC1004377B.

Bodenhorn, B. 2014. "Meeting Minds; Encountering Worlds: Science and Other Expertises on the North Slope of Alaska". In *Collaborators Collaborating*, edited by M. Konrad. Oxford: Berghahn Books.

Bourdieu, Pierre. 1977. *Outline of a Theory of Practice*. Translated by Richard Nice. Cambridge Studies in Social and Cultural Anthropology. Cambridge; New York, NY: Cambridge University Press. www.cambridge.org/core/books/outline-of-a-theory-of-practice/193A11572779B478F5BAA3E3028827D8.

Bourdieu, Pierre. 2001. *Masculine Domination*. Stanford: Stanford University Press.

Bourdieu, Pierre. 2008. *The Bachelor's Ball: The Crisis of Peasant Society in Béarn*. Translated by Richard Nice. Cambridge: Polity.

Bourdieu, Pierre, and Loïc J. D. Wacquant. 2002. *An Invitation to Reflexive Sociology*. Cambridge: Polity Press.

Bourn, Nigel A. D., and C. R. Bulman. 2005. "Landscape Scale Conservation, Theory into Practice". *Studies on the Ecology and Conservation of Butterflies in Europe* 1: 111–112.

Case, Philip. 2016. "Farmer Loses Fight for Abstraction Licence". Industry publication. *Farmers Weekly* (blog). 20 September 2016. www.fwi.co.uk/news/farmer-loses-fight-for-abstraction-licence.htm.

Clarke, Mike, Steve Ormerod, and Danae Sheehan. 2013. "RSPB Annual Review 2012–2013". Annual Report. United Kingdom: Royal Society for the Protection of Birds.

du Gay, Paul. 2000. *In Praise of Bureaucracy: Weber, Organization, Ethics*. London: SAGE.

Dunford, Daniel, and Ashley Kirk. 2016. "EU Referendum: Leave Supporters Trust Ordinary 'Common Sense' More than Academics and Experts". *Telegraph*, 2016. www.telegraph.co.uk/news/2016/06/16/eu-referendum-leave-supporters-trust-ordinary-common-sense-than/.

Ellis, Sam, Dave Wainwright, Frank Berney, Caroline Bulman, and Nigel Bourn. 2011. "Landscape-Scale Conservation in Practice: Lessons from Northern England, UK". *Journal of Insect Conservation* 15 (1–2): 69–81. https://doi.org/10.1007/s10841-010-9324-0.

Foster, Russell. 2016. "'I Want My Country Back': The Resurgence of English Nationalism". *LSE BREXIT* (blog). 6 September 2016. http://blogs.lse.ac.uk/brexit/2016/09/06/i-want-my-country-back-the-resurgence-of-english-nationalism/.

Frake, Charles. 1996. "A Church Too Far Near a Bridge Oddly Placed: The Cultural Construction of the Norfolk Countryside". In *Redefining Nature : Ecology, Culture and Domestication*, edited by R. F. Ellen and Katsuyoshi Fukui. Oxford: Berg.

García, Raúl Sánchez, and Dale C. Spencer. 2014. *Fighting Scholars: Habitus and Ethnographies of Martial Arts and Combat Sports*. London: Anthem Press.

Gasparino, Charles. 2016. "Brexit Won Because Common Sense Prevailed". *New York Post* (blog). 24 June 2016. http://nypost.com/2016/06/24/brexit-won-because-common-sense-prevailed/.

George, Martin. 2016. *The Hickling Broad Nature Reserve*. Norwich: Norfolk Wildlife Trust.

Goet, Niels. 2016. "Brexiteers Must Fall: Why Liberals and the Left Must Combine Forces to Confront the Cecil Rhodes of the Twenty-First Century". *OxPol* (blog). 8 July 2016. http://blog.politics.ox.ac.uk/brexiteers-must-fall-liberals-left-must-combine-forces-confront-cecil-rhodes-twenty-first-century/.

Green, Maia. 2016. "Anthropology and Organisational Change: Gillian Tett's The Silo Effect". *Savage Minds* (blog). 17 August 2016. http://savageminds.org/2016/08/17/anthropology-and-organisational-change-gillian-tetts-the-silo-effect/.

Hamel, Jacques. 1997. "Sociology, Common Sense, and Qualitative Methodology: The Position of Pierre Bourdieu and Alain Touraine". *The Canadian Journal of Sociology/Cahiers Canadiens de Sociologie* 22 (1): 95–112. https://doi.org/10.2307/3341565.

Hardin, Garrett. 1968. "The Tragedy of the Commons". *Science* 162 (3859): 1243–1248. https://doi.org/10.1126/science.162.3859.1243.

Harper, Martin. 2014. "The Case for Catfield". The RSPB Community. *Martin Harper's Blog* (blog). 14 July 2014. www.rspb.org.uk/community/ourwork/b/martinharper/archive/2014/07/14/water-water-not-quite-everywhere.aspx.

Harper, Martin. 2015. "The Case for Catfield (Part 3 and Conclusion?)". The RSPB Community. *Martin Harper's Blog* (blog). 15 May 2015. www.rspb.org.uk/community/ourwork/b/martinharper/archive/2015/05/15/sutton-and-catfield.aspx.

Harper, Martin. 2016. "Good News for a Friday: Congratulations to Catfield Coalition". The RSPB Community. *Martin Harper's Blog* (blog). 23 September 2016. www.rspb.org.uk/community/ourwork/b/martinharper/archive/2016/09/23/good-news-for-a-friday-congratulations-to-catfield-coalition.aspx.

Harris, Tim, Alec Bull, Keith Clarke, Peter Lambley, Trevor Dodson, Peter Nicholson, Peter Collyer, *et al.* 2008. *A Natural History of the Catfield Hall Estate*. Occasional Publication 11. Norwich: Norfolk and Norwich Naturalists Society.

Harvey, Penelope. 2007. "Introduction: Expertise, Technology and Public Culture". *Sociological Review* 55 (1): 1–4.

Holton, Robert. 2000. "Bourdieu and Common Sense". In *Pierre Bourdieu: Fieldwork in Culture*, edited by Nicholas Brown and Imre Szeman, 87–99. Lanham, MD: Rowman & Littlefield.

Kershaw, Ian, Anthony Seldon, Selina Todd, Hakim Adi, Juliet Gardiner, and Vernon Bogdanor. 2016. "David Cameron's Legacy: The Historians' Verdict". The *Guardian*, 15 July 2016, sec. Politics. www.theguardian.com/politics/2016/jul/15/david-camerons-legacy-the-historians-verdict.

Lane, Jeremy F. 2006. *Bourdieu's Politics: Problems and Possibilities*. London: Routledge.

Lizardo, Omar, and Michael Strand. 2010. "Skills, Toolkits, Contexts and Institutions: Clarifying the Relationship between Different Approaches to Cognition in Cultural Sociology". *Poetics*, Brain, Mind and Cultural Sociology, 38 (2): 205–228. https://doi.org/10.1016/j.poetic.2009.11.003.

Locke, John. 1988. *Two Treatises of Government*. Cambridge: Cambridge University Press.

MacKenzie, Donald. 2009. "All Those Arrows". *London Review of Books*, 25 June.

Macpherson, C. B. 1962. *The Political Theory of Possessive Individualism: Hobbes to Locke*. 1st edition. Oxford: Clarendon Press.

Marková, Ivana. 2016. "Towards Giambattista Vico's Common Sense". In *The Dialogical Mind: Common Sense and Ethics*, 39–61. Cambridge: Cambridge University Press. www.cambridge.org/core/books/the-dialogical-mind/towards-giambattista-vicos-common-sense/1D30A68ADE98D107593068960265B478.

McKenna, Brian. 2011. "Bestselling Anthropologist 'Predicted' Financial Meltdown of 2008". *Society for Applied Anthropology Newsletter*, 2011. http://192.163.234.187/~sfaanet/news/files/5113/7493/9950/22-1.pdf.

Mintz, Sidney Wilfred. 1986. *Sweetness and Power: Place of Sugar in Modern History*. New York, NY: Penguin Books Ltd.

Morrin, Kirsty. 2015. "Unresolved Reflections: Bourdieu, Haunting and Struggling with Ghosts". In *Bourdieu: The Next Generation: The Development of Bourdieu's Intellectual Heritage in Contemporary UK Sociology*, edited by Jenny Thatcher, Nicola Ingram, Ciaran Burke, and Jessie Abrahams. London: Routledge.

Münchau, Wolfgang. 2017. "A Warning for the Losers of the Liberal Elite". *Financial Times*, 15 January 2017. www.ft.com/content/646cf682-d9bc-11e6-944b-e7eb37a6aa8e.

Neal, Justin. 2016. "Environmental Brexit". *Aaron & Partners LLP – Blog* (blog). 20 September 2016. www.aaronandpartners.com/environmental-brexit/.

Norfolk Wildlife Trust. 2017. "Hickling Broad Land Purchase Appeal – Norfolk Wildlife Trust". Norfolk Wildlife Trust. 2 April 2017. www.norfolkwildlifetrust.org.uk/support-us/hickling-broad-land-purchase-appeal.

O'Riordan, Tim, and Rosie Ferguson. 2000. "Seeking Reconciliation for Hickling and the Broads Generally – Preliminary Report," 21 June.

Ostrom, Elinor. 1991. *Governing the Commons: The Evolution of Institutions for Collective Action*. Cambridge; New York, NY: Cambridge University Press.

Ostrom, Elinor. 2003. "Toward a Behavioural Theory Linking Trust, Reciprocity and Reputation". In *Trust and Reciprocity: Interdisciplinary Lessons from Experimental Research*, edited by Elinor Ostrom and J. Walker. New York, NY: Russell Sage Foundation.

Ostrom, Elinor. 2009. *Understanding Institutional Diversity*. Princeton: Princeton University Press.

Ostrom, Elinor, and J. Walker. 2003. "Introduction". In *Trust and Reciprocity: Interdisciplinary Lessons from Experimental Research*, edited by Elinor Ostrom and J. Walker. New York, NY: Russell Sage Foundation.

Page, Ben, and Karen Bakker. 2005. "Water Governance and Water Users in a Privatised Water Industry: Participation in Policy-Making and in Water Services Provision. A Case Study of England and Wales". *International Journal of Water* 3 (1). https://doi.org/10.1504/IJW.2005.007158.

Parker, George, Michael Mackenzie, and Ben Hall. 2016. "Britain Turns Its Back on Europe". *Financial Times*, 24 June 2016. www.ft.com/content/e404c2fc-3913-11e6-9a05-82a9b15a8ee7.

Paxman, J. 2001. *The English: A Portrait of a People*. London: Penguin.

Penny, Laurie. 2016. "I Want My Country Back". *New Statesman*, 2016. www.newstatesman.com/politics/uk/2016/06/i-want-my-country-back.

Poole, Steven. 2015. "The Silo Effect by Gillian Tett Review – a Subversive Manifesto". *Guardian*, 17 October 2015, sec. Books. www.theguardian.com/books/2015/oct/17/the-silo-effect-why-putting-everything-in-its-place-isnt-such-a-bright-idea-gillian-tett-review.

Rothstein, Bo. 2000. "Trust, Social Dilemmas and Collective Memories". *Journal of Theoretical Politics* 12 (4): 477–501. https://doi.org/10.1177/0951692800012004007.

Shiach, Morag. 1993. "'Cultural Studies' and the Work of Pierre Bourdieu". *French Cultural Studies* 4 (12): 213–223. https://doi.org/10.1177/095715589300401203.

Strang, Veronica. 2004. *The Meaning of Water*. Oxford; New York, NY: Bloomsbury Academic.

Swinkels, Michiel, and Anouk de Koning. 2016. "Introduction: Humour and Anthropology". *Etnofoor* 28 (1): 7–10.

Tett, Gillian. 2016. *The Silo Effect: Why Every Organisation Needs to Disrupt Itself to Survive*. London: Abacus.

The National Farmers' Union. 2015. "Q&As – the EU". NFU Online. 21 October 2015. www.nfuonline.com/news/eu-referendum/qa-the-eu/.

Tsing, Anna. 2015. *The Mushroom at the End of the World: On the Possibility of Life in Capitalist Ruins.* Princeton: Princeton University Press.

Tully, James. 1980. *A Discourse on Property: John Locke and His Adversaries.* Cambridge: Cambridge University Press. http://ebooks.cambridge.org/ref/id/CBO9780511558641.

Van Reenen, John. 2016. "The Aftermath of the Brexit Vote – the Verdict from a Derided Expert". *British Politics and Policy at LSE* (blog). 2 August 2016. http://blogs.lse.ac.uk/politicsandpolicy/the-aftermath-of-the-brexit-vote-a-verdict-from-those-of-those-experts-were-not-supposed-to-listen-to/.

Vico, Giambattista. 1948. *The New Science of Giambattista Vico.* Translated by Thomas Goddard Bergin and Max Harold Fisch. Ithaca, NY: Cornell University Press. https://archive.org/stream/newscienceofgiam030174mbp/newscienceofgiam030174mbp_djvu.txt.

Wacquant, Loïc. 2001. "Critical Thought as Solvent of Doxa". *European Institute for Progressive Cultural Politics* (blog). 2001. http://eipcp.net/transversal/0806/wacquant/en.

Wheeler, Brian, and Alex Hunt. 2016. "Brexit: All You Need to Know about the UK Leaving the EU". *BBC News,* 2 October 2016, sec. Brexit. www.bbc.co.uk/news/uk-politics-32810887.

Williams, Zoe. 2016. "Brexit Fallout: Six Practical Ways to Help Fix This Mess". *Guardian,* 28 June 2016, sec. Politics. www.theguardian.com/politics/2016/jun/28/brexit-fallout-six-practical-ways-to-fix-this-mess.

Williamson, Tom. 1997. *The Norfolk Broads: A Landscape History.* 1st edition. Manchester, UK; New York, NY: Manchester University Press.

Wood, Neal. 1984. *John Locke and Agrarian Capitalism.* Berkeley: University of California Press.

Woolley, Jonathan. 2018. "Rede of Reeds: Land and Labour in Rural Norfolk". PhD Dissertation. Cambridge: University of Cambridge. www.repository.cam.ac.uk/handle/1810/273374.

Conclusions

What do we need to know about common sense?

Man is but a reed, the most feeble thing in nature; but he is a thinking reed.
Blaise Pascal, *Pensées* – n347 (2014 [1688])

Scholars may treat common sense as a matter of reason or received wisdom; but studying the phrase ethnographically in vernacular English reveals a quite different and culturally particular set of meanings. First and foremost, common sense refers to the correct attitude for a shared working environment – what Tim Ingold might call a "taskscape" – in short, a commons. A common sense attitude is deemed correct in that it supports both morally upright and practically useful behaviour. It isn't consciously taught, but developed organically by sharing common tasks with others. In other words, common sense is acquired by developing a shared sense of your collective practices – a habitus of the common – with fellow commoners. So powerful is the common, that for those whose habitus is quite distinct – now the majority of the population – the common acts as a root metaphor, drawing together widely held ideas about society, knowledge, and the landscape. But because the common still holds this power, faith in common sense raises false hopes of easy consensus when shared resources are put under pressure, and objective crises occur. And when consensus is frustrated by different, conflicting interests, expertise, and experience, the moral dimension of common sense encourages English people to transmute that frustration into mistrust and enmity. *Rather than put time, effort, and hard cash into maintaining common ground, the English tend not to bother – assuming their particular views are already sensible to everyone.* The false hope in an easy common sense ensures that it is now siloing, and not commoning, that dominates the English landscape. All that remains is to reflect: what bearing does this failure of common sense have on the relationship between human beings and the wider landscape?

One of the most iconic symbols of environmental land management in Broadland is the reed (genus *Phragmites*). As described above, reedbed and fen are emblematic of the broads at their well-managed best; these are habitats that swiftly disappear without intervention, due to succession, and the damage to Broadland in the twentieth century was partly manifested through the loss of reedy shallows at the edges of the watercourses. Although I have not chosen to follow reeds as "actants" in the fashion of actor-network theorists (Latour 1993; Goodman and Walsh 2001), they are nonetheless a constant presence in my ethnography – being cut in rotation on Strumpshaw Fen, growing along the paths I walked, and roofing the houses of many of my informants. For Pascal, in the quote with which this conclusion begins, these reeds symbolise human frailty. Human beings are so easily slain; what ennobles us as a species is not our bodily existence, but our thoughts: "It is not from space that I must seek my dignity, but from the government of my thought.... By space the universe encompasses and swallows me up like an atom; by thought I comprehend the world" (Pascal 2014 [1688], n348). In this classic Enlightenment refrain, human reason is both the highest accomplishment to which we might aspire, and a cavernous gulf between man and all else in creation. The sole counsel a reed can provide is as a metaphor of frailty.

But it is possible to take the metaphorical juxtaposition of man and reed in an altogether different direction. As the sage Aesop once pointed out – in a story that has become a folktale in English culture – while an oak tree trusts in its strength and is blown down in a gale, the reed bends in the wind and survives. There is a very sound evolutionary logic to being flexible. It allows you to adapt to a wide variety of niches, across many different habitats; when any one of those niches disappears, you can simply find another. Humans possess a suite of physical and cognitive features that equip us to be adaptable in this way, a central feature of which is a core concern of anthropology: *culture*. Culture can be acquired, changed, and dispensed far more rapidly than most physical traits. Like the reed, humans bend. But contrary to Pascal's grim pronouncements, our willingness to feebly go along with whatever niches pass by is a source of our strength as a species, as much as it is a weakness. Pascal and his contemporaries – so seduced were they by the Classical eidolon of transcendent, universal Reason, sundered from base Nature – that they overlooked the power that comes with the naturalness of the human experience. Selective pressure has driven our species toward cultural flexibility rather than the doughty physical constitutions of other animals. Why incubate an inbuilt resistance to poisons, if you can take antitoxins from plants to treat them as and when? Why grow thick fur, if you can take it from other animals? Why invest in claws, when you can figure out how to make blades of stone, or bronze, or iron, or carbon fibre?

This point has direct relevance to the management of the Broads. In his stirring polemic, *Feral* (2013), journalist George Monbiot suggests that the ability of certain British tree species – like hazel (*Corylus avellana*), ash (*Fraxinus excelsior*), oak (*Quercus robur*), beech (*Fagus sylvaticus*), alder (*Alnus glutinosa*), and willow (genus *Salix*) – to coppice, or re-sprout after their trunk is snapped, is evidence of the browsing of forest elephants during the Eemian interglacial 131–114 kya (Monbiot 2013, 90–91).[1] Although Monbiot's main point is to cite this as evidence for how far the baselines of mainstream conservation have shifted in Britain – we don't seek to re-introduce elephants to the countryside – I would suggest that this observation has far wider implications. The ability of trees to coppice, although originally an adaptation to the predation of elephants, has been harnessed by humans as a way of sustainably harvesting wood for fuel, shelter, and tool manufacture. Humans have – in a sense – taken on an ecological niche originally filled by another animal, in a sense "becoming" forest elephants. This same process can be observed in a vast array of different tasks that humans undertake in the English landscape. When we mow hay meadows, cut down reeds, or deedle (dig) dykes, we are performing the role of aurochs (*Bos primigenius*) or wisent (*Bison bonasus*); when we kill foxes (*Vulpes vulpes*) or crows (*Corvus corone*) we are acting as wolves (*Canis lupus*) or eagles (genus *Aquila*); when we build mill ponds and weirs we act as beavers (*Castor fiber*). Even the humble robin (*Erithacus rubecula*) that is beloved by British gardeners for following them closely as they work is just doing this because gardeners, in disturbing the earth's invertebrates with their digging, are behaving like wild boar (*Sus scrofa*), which robins evolved alongside. Through our adaptability, humans are capable of occupying the niches of all manner of other animals, moving between them depending upon the circumstances. As Marshall Sahlins points out, "Tools, even good tools, are prehuman. The great evolutionary divide is in the relationship: tool-organism" (Sahlins 2003, 80). But the real difference, perhaps, is not the tenor of that relationship, but in the number of relationships one can forge. We are the universal keystone species, culture engaging us in a constant process of becoming (Deleuze and Guattari 1987, 256–352). Equipped with a cultural propensity to adapt, human beings can hop from niche to niche – the original silo-busters.

This alternative metaphorical engagement with humanity and reeds reveals a deeper, non-metaphorical point: that humans are, and have always been, enmeshed within the landscapes of which we are only part. We are not merely a thinking reed; but reeds – like all else in nature – are, in the words of Claude Levi-Strauss, good to think with (Levi-Strauss 1962). Indeed, the world in which we live is the very substance of our thoughts. Within an English context, common sense – in one of its vernacular guises as a pragmatic, situationally rational attitude – is an attempt to embrace this ecologically-including-socially-situated flexibility. Common sense's very particularity, its rootedness in particular working environments, makes it localised, but its nature as an attitude toward life in general makes it mobile. Anthropologists have argued for decades about what defines our species – are we *Homo economicus*, *Homo faber*, or *Homo narrans*? I would contend that the nature of human beings has been under our very noses the entire time: the very nature of humankind is to step outside of culture, while never being able to fully escape it – to be insiders, and outsiders, of each niche we occupy. The practice of anthropology itself is a refinement of the essence of our species – being to bend in the winds of culture, but to simultaneously reflect upon it, to transcend it, and to spin the webs of meaning anew. Anthropology is itself silo-busting in the same sort of way – it allows for human beings to think themselves out of their established cultural frames (see Figure C.1). Common sense aspires to this same virtue; but its particularity – in our increasingly complex, specialised world – frustrates its capacity to do so.

Gillian Tett, Robert Kett, and the division of labour

Though she does not mention his name, the process of siloing highlighted by Gillian Tett in modern societies has dramatic consequences for the theories of Emile Durkheim, a founding father of French sociology. His concept of organic solidarity – in which modern societies sustain cohesion through the economic interdependence of highly differentiated individuals (Durkheim 2014, 102), predicated on the "necessary interaction of units" (Shanin 1971, 14) – is logical. The division of labour does indeed make collaboration vital. However, just because that collaboration may be deemed objectively vital does not ensure that the individual people – or indeed, groups of people – will necessarily feel moved to collaborate, or have common views about the terms of that collaboration. There is a large gap here, between what will and what ought. Different common senses, particular to specific workplaces due to their habitual particularity, resist standardisation and frustrate organic solidarity. Rather than repeat the divisions struck by the structure and agency debate, what we see here is a conflict between different heterogeneous,

Figure C.1 Thinking reeds. The reed-like flexibility of human culture allows our species to adopt numerous ecological niches.

localised structures – a patchwork of habitus. While this does not cause issues where commonable resources are managed exclusively as a single workplace, this is not the case in the modern landscape. The division of labour, rather than serving as the ultimate force for social cohesion, because it also entails a division of *understanding*, has the opposite effect to what Durkheim predicted – an effect, incidentally, to which English society, with its shared aspiration for a community based on collective labour, is particularly vulnerable. Exasperated with the effects of siloing, English people lose faith in society, and withdraw from civic life – inadvertently making alienation even worse, leading to what Durkheim referred to as the "anomic division of labour" (Durkheim 2014, 277–290). Bureaucracy, with its clearly defined roles and vocational spirit (Weber 2003), is to the world of work what siloing – with its entrenched, inflexible thinking – is to sociality (Woolley 2018, chapter 5). Durkheim's theories have lent credence to a world of human individuals as cogs in a gloriously orderly machine, each with clear functions and the labour divided between them. This mechanistic vision diminishes humanity's greatest adaptation – adaptability – and has laid waste many of the ecosystems that selected it. The possessive individuals of Norfolk are reeds no longer; they are all oak trees now.

This insight is significant given the profound influence that Durkheimian sociology has had – and continues to have – with respect to English society. For the English, common sense is not just perceived to be an expression of social unity; it is deemed to be a precondition for social cohesion. Given that it emerges from working practices, as those practices become more specialised and divided up according to progressively more focused kinds of professional expertise, the less any truly common sense can be found. The sprawling bureaucracies of England today, the state agencies and corporations that they compose, circulating flows of information and resources, rely upon an intuitive, unstated assumption of the validity of the concept of organic solidarity. By treating social cohesion – working together, building consensus, and sharing information – as a natural or necessary feature of highly networked social institutions, rather than an aspiration that needs to be constantly and actively pursued, such organisations are setting in store numerous problems, as Tett identified, due to tunnel vision and an incapacity to correctly identify both risk and opportunity. Such problems, I suggest, have become progressively compounded over the course of the past century, culminating in catastrophic events that have reshaped the political economy of the whole world – from the Financial Crisis of 2008, to the vote to leave the European Union in 2016.

My fieldwork in the Broads also suggests that we must go further than Tett and cultivate an awareness of how the division of labour affects not just human relationships, but the relationships between humans and their wider landscape. Robert Kett's demands – rooted in rights to fish and pasture – are a case in point of how important this connection is, how questions of working together in groups are never that far away from questions of how we work with the resources we have at our disposal. The division of labour does not just preclude social solidarity; but, as we have seen, it creates the conditions for the majority of human beings to become economically alienated from the landscape of which they are an undeniable part. Primary industries like agriculture, aquaculture, and forestry are professionalised, personalised, and mechanised in the Broads today, with the result that even those who live in these sectors are seeing their relationship with the land transform radically. And for the majority of people, the Broads is simply one destination among many: a place one goes to, not a place to which one belongs – a commodity you can take or leave. In response to the removal of words like *acorn*, *bluebell*, *cygnet*, *hazel*, *mistletoe*, *otter*, and *pasture* from the Oxford Junior Dictionary, in favour of ones like *block-graph*, *broadband*, *committee*, and *cut-and-paste*, Robert Macfarlane laments that "a basic literacy of landscape is falling away up and down the ages. A common language – a language *of* the commons – is getting rarer" (Macfarlane 2015, 3–4, original emphasis). What Macfarlane observes in language, I suggest, are lexical traces of a broader ecopathology – a gradual dissolution of English people's intimate familiarity with the substance of the landscape, a rift that is practical and intellectual in nature. Late capitalism, with its mechanistic, abstract logics, has imposed upon humanity a series of entrenched silos that threaten our ability to respond to the changing circumstances in which we find ourselves (see Figure C.2).

And while many people's interest and labour has been diverted elsewhere, the impact of their inhabitation has not. Carbon dioxide, treated sewage, agricultural runoff, and urban sprawl radiate out from the human parts of Broadland, all with their concomitant impact upon the local flora and fauna. This process of enclosure and alienation, I suggest, is a major reason why English people continue to disagree about climate change and other ecological issues – the majority simply aren't paying attention. Stuck in the silo of a service-sector-dominated economy, separated from the land by their specialised labour practices, they simply don't pay much attention to the way their country is changing all around them. As Tett warns, such blinkered tunnel vision can lead people to ignore catastrophic risks until it's far too late. For the Broads, with its tidal rivers and the low-lying topography, one of the parts of the British Isles most acutely vulnerable to climate change and rising sea levels, this is a topic of acute concern to conservationists and local officials (Woolley 2016). And despite the fact that official figures predict that the seas will have risen by 37 cm by 2080, putting properties in the lower reaches at risk of flood damage (Conti and Long 2011, 19), riverfront properties are continuing to fetch high prices on the local housing market. It is deeply ironic that a social order that should exemplify a state of *organic* solidarity should have the ultimate effect of being so utterly destructive to all living parts of the Earth system. The same socio-economic process that has encouraged the people of Norfolk to become oaks, has raised up a storm.

Figure C.2 Oak trees killed by rising water levels in Cantley Marsh. The division of labour encourages specialisation and inflexibility; in the Anthropocene, such traits are highly maladaptive.

Note

1 Monbiot first heard this idea suggested by a forester named Adam Thorogood, and identified the theory in a paper by Oliver Rackham, who may have originated it (Rackham 2002, 3).

References

Conti, Maria, and Andrea Long. 2011. "The Broads Plan 2011". Strategic Plan. Broads Plan. Norfolk: Broads Authority. www.broads-authority.gov.uk/__data/assets/pdf_file/0015/402045/Broads-Plan-2011.pdf.

Deleuze, G., and F. Guattari. 1987. *A Thousand Plateaus*. Translated by B. Massumi. London: Continuum.

Durkheim, Émile. 2014. *The Division of Labor in Society*. Translated by W. D. Halls. New York: Free Press.
Goodman, Jordan, and Vivien Walsh. 2001. *The Story of Taxol: Nature and Politics in the Pursuit of an Anti-Cancer Drug*. 1st edition. Cambridge; New York, NY: Cambridge University Press.
Latour, Bruno. 1993. *The Pasteurization of France*. Translated by Alan Sheridan and John Law. Cambridge, MA: Harvard University Press.
Levi-Strauss, C. 1962. *Totemism*. London: Merlin Press.
Macfarlane, Robert. 2015. *Landmarks*. 1st edition. London: Hamish Hamilton.
Monbiot, George. 2013. *Feral: Searching for Enchantment on the Frontiers of Rewilding*. London: Penguin.
Pascal, Blaise. 2014. *Pensées*. Translated by W. F. Trotter. Adelaide: The University of Adelaide Library.
Rackham, Oliver. 2002. "Ancient Forestry Practices". In *The Role of Food, Agriculture, Forestry and Fisheries in Human Nutrition*, edited by Victor R. Squires. Vol. 2. Encyclopedia of Life Support Systems.
Sahlins, Marshall. 2003. *Stone Age Economics*. 2nd edition. London u.a.: Routledge.
Shanin, Teodor. 1971. "Introduction". In *Peasants and Peasant Societies*, edited by Teodor Shanin, 11–21. London: Penguin Books.
Weber, Max. 2003. *Political Writings/Weber*. Cambridge: Cambridge University Press.
Woolley, Jonathan. 2016. "Fieldwork in Lyonesse: Salvage Ethnography before the Anthropocene Floods". *King's Review*, 2016. http://kingsreview.co.uk/articles/fieldwork-lyonesse-salvage-ethnography-anthropocene-floods/.
Woolley, Jonathan. 2018. "Rede of Reeds: Land and Labour in Rural Norfolk". PhD Dissertation. Cambridge: University of Cambridge. www.repository.cam.ac.uk/handle/1810/273374.

Index

Page numbers in **bold** denote tables, those in *italics* denote figures.